Fuzz

FUZZ

MARY ROACH

THORNDIKE PRESS
A part of Gale, a Cengage Company

GALE
A Cengage Company

LIBRARY OF CONGRESS CIP DATA ON FILE.
CATALOGUING IN PUBLICATION FOR THIS BOOK
IS AVAILABLE FROM THE LIBRARY OF CONGRESS.

ISBN-13: 978-1-4328-9426-9 (hardcover alk. paper)

Published in 2022 by arrangement with W.W. Norton & Company, Inc.

Printed in Mexico
Print Number: 01   Print Year: 2022

For Gus, Bean, and Winnie.
To the farthest star.

CONTENTS

A QUICK WORD OF INTRODUCTION

On June 26, 1659, a representative from five towns in a province of northern Italy initiated legal proceedings against caterpillars. The local specimens, went the complaint, were trespassing and pilfering from people's gardens and orchards. A summons was issued and five copies made and nailed to trees in forests adjacent to each town. The caterpillars were ordered to appear in court on the twenty-eighth of June, at a specified hour, where they would be assigned legal representation.

Of course no caterpillars appeared at the appointed time, but the case went forward anyway. In a surviving document, the court recognizes the rights of caterpillars to live freely and happily, provided this does not "impair the happiness of man . . ." The judge decreed that the caterpillars be assigned a plot of alternate land for their sustenance and enjoyment. By the time the details were worked out, the defendants, having pupated, were surely through with their devastations, and all parties no doubt left the proceedings satisfied.

The case is detailed in an unusual 1906 book, *The Criminal Prosecution and Capital Punishment of Animals.* When I first paged through it, I wondered if it might be an ambitious hoax. Here were bears formally excommunicated from the Church. Slugs given three warnings to stop nettling farmers, under penalty of "smiting." But the author, a respected historian and linguist, quickly wore me down with a depth of detail gleaned from original documents, nineteen of which are reproduced in their original languages in a series of appendices. We have the itemized expense report of a French bailiff, submitted in 1403 following the murder trial of a pig ("cost of keeping her in jail, six sols parisis"). We have writs of ejectment issued to rats and thrust into their burrows. From a 1545 complaint brought by vintners against a species of greenish weevil, we have not only the names of the lawyers but early examples of that time-honored legal tactic, the stall. As far as I could tell, the proceedings dragged on eight or nine months — in any case, longer than the life span of a weevil.

I present all this not as evidence of the silliness of bygone legal systems but as evidence of the intractable nature of human-wildlife conflict — as it is known today by those who grapple with it professionally. The question has defied satisfactory resolution for centuries: What is the proper course when nature breaks laws intended for people?

The actions of the magistrates and prelates made no rational sense, of course, for rats and weevils cannot understand property law or be expected to conform to the moral principles of human civilizations. The aim was to cow and impress the populace: *look here, even nature must bend to our rule!* And it was, in its way, impressive. The sixteenth-century judge who granted leniency to moles with young offspring made a show not only of his authority but of his temperance and compassion.

Wandering through the Middle Ages and the centuries just beyond, I began to wonder what the modern epoch had brought to bear on these matters. Having sampled the esoteric solutions of law and religion, I set out to see what science has been bringing to the table, and what answers it might offer for the future. So began more wandering. My guides were people with titles unfamiliar to me: Human-Elephant Conflict Specialist, Bear Manager, Danger-Tree Faller-Blaster. I spent time with predator attack specialists and attack forensics investigators, builders of laser scarecrows and testers of kinder poisons. I traveled to some of the "hot spots" — back alleys in Aspen, Colorado; leopard-terrorized hamlets in the Indian Himalaya; St. Peter's Square the night before the pope's Easter Mass. I considered the contributions of bygone professionals — the economic ornithologists and the rat searchers — as well as the stewards of the future, the conservation geneticists. I taste-tested rat bait. I was mugged by a macaque.

The book is far from comprehensive. Two thousand species in two hundred countries regularly commit acts that put them at odds with humans. Each conflict needs a resolution unique to the setting, the species, the stakes, the stakeholders. What you have here is the highlights of a two-year exploration, a journey through a world I had not known existed.

The first half of the book considers the felony crimes. Murder and manslaughter, serial killing, aggravated assault. Robbery and home invasion. Body snatching. Grand theft, sunflower seed. The perpetrators include the usual suspects, the bears and the big cats, and some less usual — monkeys, blackbirds, Douglas firs. The later pages explore acts less grievous but more widespread. We consider the jaywalking ungulates. The vultures and gulls that vandalize property for no discernible reason. The littering geese and the trespassing rodents.

Of course, these are not literal criminal acts. Animals don't follow laws, they follow instincts. Almost without exception, the wildlife in these pages are simply animals doing what animals do: feeding, shitting, setting up a home, defending themselves or their young. They just happen to be doing these things to, or on, a human, or that human's home or crops. Nonetheless the conflicts exist, creating dilemmas for people and municipalities, hardships for wildlife, and material for someone else's unusual book.

1
Maul Cops: Crime Scene Forensics When the Killer Isn't Human

For most of the past century, your odds of being killed by a cougar were about the same as your odds of being killed by a filing cabinet. Snowplows kill twice as many Canadians as grizzly bears do. In the extremely uncommon instance when a North American human is killed by a wild North American mammal, the investigation falls to officers and wardens with state or provincial departments of fish and game (or fish and wildlife, as less hunty states like mine have rebranded themselves). Because the incidents are so rare, few of these men and women have much experience with them. They're more accustomed to poaching cases. When the tables turn and the animal is the suspect, a different kind of forensics and crime-scene know-how is called for.

Without it, mistakes are made. In 1995, a cougar was presumed to have killed a young man found dead on a trail with puncture wounds to the neck, while the true murderer, a human being, walked free. In 2015, a wolf was wrongfully accused of pulling a man from his sleeping bag and killing him. Cases like these are one reason there is WHART: Wildlife-Human Attack Response Training (and by its founders' admission, "a horrible acronym"). WHART is a five-day course — part lecture and part field training — taught by members of the British Columbia Conservation Officer Service.*

Because they have the experience. British Columbia has more cougar attacks than any other North American state or province. It has 150,000 black bears — to Alaska's 100,000 — 17,000 grizzlies, and 60 predator attack specialists, 14 of whom (the specialists but not the bears) have driven down from Canada to serve as WHART instructors this week. WHART 2018 is being hosted by the Nevada Department of Wildlife, which has offices in Reno. This fact helps explain why a training course for wilderness professionals would be held in a casino complex, where the resident wildlife amounts to the furry hominid on the Betti the Yetti slot machine and an unspecified

* Canadian for "Fish and Game Department."

"biohazard" that closed down the pool for a day. WHART seems to be the only booking at the Boomtown Casino event and conference center this week. Management has a bingo game going on in the next room.

The WHART student body, some eighty of us in all, has been split into small groups, each led by one of the predator attack specialists. Like many Canadians, they are distinguishable from white Americans mainly by sound. I'm referring to that uniquely far-northern habit of ending statements with folksy interrogatives. It's an endearing custom thrown somewhat off-kilter by the present subject matter. "Quite a bit of consumption and feedin' and whatnot, eh?" "Holdin' on by two, three tendons, right, ya know?"

Our conference room, the Ponderosa, is a standard offering with a podium and a screen for slides and videos. Less standard are the five large animal skulls sitting in a row on a long table at the front of the room, like participants in a panel discussion. On the screen, a grizzly bear is attacking Wilf Lloyd of Cranbrook, British Columbia. The footage is part of a presentation entitled "Tactical Killing of a Predator on a Person." The instructor sums up the challenge that Wilf's son-in-law faced in trying to shoot the bear but not the man: "All you could see was the body of the bear and a limb of Wilf once in a while." The son-in-law saved Wilf's life but also shot him in the leg.

Another challenge: Marksmanship deteriorates under the influence of adrenaline. Fine motor skills are out the window. The thing to do, we are told, is to "run directly up to that animal, plant the barrel and shoot upward" to avoid hitting the victim. Though you then run the risk of "attack redirection." That's a calm, technical way to say that the animal has dropped its victim and now it's coming after you.

A second video illustrates the importance of order and discipline in the face of animal-attack mayhem. In it, a male lion charges a safari hunter. The other members of the hunting party wheel and scatter. The video is paused at various moments when a rifle is pointing both at the lion and at a hunter directly behind it. "Stay tight and communicate," is the advice here. We will be practicing this kind of thing later, in an immersive field scenario out in the scrub near the Truckee River, below the casino.

The cursor glides to the Play arrow again, and the lion resumes its charge. I used to work at a zoo, and the roaring in the Lion House at feeding time was God-like. It twisted my viscera. And that was just their mealtime conversation. The lion in this video means to intimidate and destroy. The bingo party has to be wondering what the hell is going on in the Ponderosa Room.

After one more presentation, we break for lunch. Preordered sandwiches are waiting for us to pick up at a small deli over in the casino. We stand in line, attracting curious glances. It's unusual, I suppose, to see so many uniformed law enforcement professionals inside a gambling establishment. I collect my lunch sack and follow along behind a small group of conservation officers heading to the lawn outside. Their leather hiking boots squeak as they walk. "So she looks in her rearview mirror," one is saying, "and there's a bear in the back seat, eating popcorn." When wildlife officers gather at a conference, the shop talk is outstanding. Last night I stepped onto the elevator as a man was saying, "Ever tase an elk?"

While we were off on lunch break, the instructors stacked the chairs against the walls and laid out soft-touch male and female training manikins on the tables, one per group. Working from photographs, some of the more artistically inclined instructors have used paint and, apparently, hacksaws to create convincing facsimiles of actual wounds from attacks. *Wounds* is a tepid word for what teeth and claws can do.

My group's manikin is a female, though it would be difficult to know this from what remains of her face, or from the sign attached to the table, which reads BUD. Later, walking to the bathroom, I pass a badly mauled LABATT and a decapitated MOLSON. Instead of being numbered, the manikin workstations have been beered. I take this to be an effort, a very Canadian-dude effort, to lighten the mood.

Our first task is to apply our newly acquired forensics savvy and determine what species it was that did the mauling. We're looking at what's known in attack forensics as "victim evidence": injuries and clothing. The worst of the visible damage is above our manikin's shoulder. (Only one remains.) Part of her neck is flayed, and a flap of scalp hangs loosened, like peeling stucco. Missing eyelid, nose, lips. We all agree it doesn't seem like the work of *Homo sapiens.* Humans rarely eat their victims. If a murderer removes body parts, it's likely to be hands or head — to stymie matches with fingerprints or dental records. Murderers occasionally take a trophy, but a shoulder or lip would be an unusual choice.

The consensus is that she was killed by a bear. Bears' teeth are their main weapon, and their lightly furred face is their weak spot. When bears attack humans, they apply the tactics they use in fights with other bears. "They go teeth to teeth, right? So their instinct is to go right for your face." Joel Kline, our youthful, forthright instructor, has been an investigator on ten cases of bear attack. "They come right at you and you have all these massive injuries right to the face." Joel's own face — our focus as we take in his words — is blue-eyed, unblemished, peachy clear. I work hard not to picture it in that state.

Bears are inelegant killers partly because they're omnivores. They don't regularly kill to eat, and evolution has equipped them accordingly. They feed on nuts, berries, fruit, grasses. They scavenge trash and carrion. A cougar, by contrast, is a true carnivore. It lives by the flesh of animals it kills, and thus it kills efficiently. Cougars stalk, well hidden, and then pounce from behind and deliver a "killing bite" to the back of the neck. Their molars close like scissors blades, cutting flesh cleanly. A bear's mouth evolved for crushing and grinding, with flat molar surfaces and jaws that move side to side as well as up and down. Wounds made by bears' teeth are cruder.

And more numerous. "Bears are more bite bite bite bite." Our manikin, says Joel, is how it usually goes. "It's a big mess."

Looking around at the manikins, I see not just bites and scratches but broad scalpings and skinnings. Joel explains the mechanics of this. A human skull is too large and round for a bear or cougar to position between its jaws and get the leverage it would need to crush or bite into it. So when it brings its teeth together, they may skid off the skull and tear away skin. Think of biting into a very ripe plum, how the skin pulls away.

Deer, a popular entrée among cougars, have longer, more muscled necks than we have. When a cougar tries to make its trademark killing bite on a human, its teeth may encounter bone where normally there would be muscle. "They try to dig their canines in and they bring their teeth together and they take the flesh and remove it," said WHART co-founder Kevin Van Damme, in a talk called "Cougar Attack Behavior." Van Damme has astronaut looks and a voice that carries to the back of the Ponderosa Room without a microphone. I opened a decibel meter app on my phone at one point and was impressed to see him hit

79, about the level of a garbage disposal.

The claw marks on our simulated victim rule out a cougar. Cats' claws, unlike dogs', create a cluster of triangular punctures as they sink in to grip their prey. With a bear attack, you're more likely to see what we have here in front of us, the parallel rakings of a swipe.

Joel takes a step closer to the manikin's head. " 'Kay, what else do we have here? Missing nose, lips, right? So later we're going to think of looking for those in . . . ?"

"The bear's stomach," a few of my group mates call out.

"Stomach contents,* right on." Joel says

* Scientists with the long-ago Division of Economic Ornithology used stomach contents as evidence in cases of birds accused of raiding farms, hunting stock, and commercial fishing operations. A 1936 U.S. Department of Agriculture report provides examples: eiders accused of decimating scallop beds, yellow-crowned night herons shot by froggers when in fact the birds had been eating crayfish, hunters killing marsh hawks because they thought they were preying on quail. In each case, the birds were exonerated by their stomach contents, a happy outcome for all except of course the individuals examined, who gave their stomachs that others might live. Maryland's

"Right on" a lot. Writing the chapter later, I would recall "bingo"s, too, but that may be a memory that seeped in from the other side of the wall.

None of the manikin torsos in the room are laid open. There's none of what Van Damme calls "feeding on innards." I'm initially surprised by this. I know from research for a previous book that predatory carnivores tend to tear into the abdomen of their prey straightaway to get to the organs — the most nutritious parts. One possible reason you don't see this as much on human victims, say our instructors, is that humans wear clothing. Both bears and cougars avoid clothed areas when they're feeding or scavenging. Perhaps they don't like how the cloth feels or tastes, or they don't realize there is meat underneath.

Joel indicates a suite of wounds on the neck and shoulder. "Are we thinking perimortem or postmortem?" In other words, was our victim alive or dead as these wounds were inflicted? It's important to know this, because otherwise a bear that was just

---

Patuxent Wildlife Research Center housed a collection of thousands of glass jars of bird stomach contents, until pressing storage needs triggered a massive emesis into a Patuxent dumpster.

scavenging could take the fall for a killing. Based on the bruising around the puncture wounds, we judge them to be perimortem. Dead people don't bleed or bruise, a bruise being essentially a bleed beneath the surface of the skin. If blood is not being pumped, it doesn't flow.

Joel tells us the story of a gnawed-upon corpse that was found near its car in the woods, partially buried under leaves. The bites appeared to have come from a bear, and a bear was trapped nearby, but there was little blood on and around the man's body. Investigators found needle marks between the toes and a used syringe on the car floor. An autopsy confirmed that the man had died of an overdose. The bear, as Joel says, "just saw an opportunity to get some good, high fat and calorie content" and pulled him from the car and ate some of him and cached the body to come back to later. The bear was released.

Joel rolls our manikin onto its front side, revealing one or two additional perimortem gashes on the back. I point out two small divets along the spine, which exhibit no purpling or blood. I hazard a guess, based on a slide from yesterday showing postmortem rodent damage, that a small woodland creature might have been gnawing on our corpse. Joel exchanges a look with one of my group mates, a wildlife biologist from Colorado.

"Mary, those are marks from the injection molding." Part of the manufacturing process of the manikin, he means. This would be less embarrassing for me had I not, as group notetaker in an earlier exercise, transcribed teeth-wound measurements using the abbreviation for centimeters instead of millimeters, entering into evidence a tip-to-tip canine-tooth span not seen since the Jurassic period.

We move on now from victim evidence to animal evidence: evidence on or in a "suspect" that has been shot or captured near the scene of the attack. For instance, Joel is saying, you can look for the victim's flesh up in the pockets of the gums of the (immobilized) animal. It's odd to think of a bear getting human stuck between its teeth, but there you go.

With cougars, Joel adds, it's sometimes possible to recover the victim's blood or flesh from the crevice on the interior of a claw. "So you need to push those out, those retractable claws, and you might have evidence under there, right?"

Claws can be misleading as indicators of the size of an attacker's paw. When the animal steps down and transfers its weight onto a foot, the toes splay, making the foot appear larger. Investigators have to be cautious with measurements of claw or tooth holes in clothing as well, because the cloth could have been wrinkled or folded over as it was pierced.

" 'Kay, what else are we looking for?"

"Victim's blood on the fur?" someone offers.

"Yup, right on." Joel cautions that if the bear had been shot at the scene of the attack (rather than trapped afterward), its blood could mingle with the victim's blood and muddy the DNA tests. "And how do we prevent that?"

"Plug the wound!" And that is why men with the British Columbia Conservation Officer Service keep a box of tampons in the truck.

What we're seeking, the end point of all this, is linkage: crime-scene evidence that connects the killer to the victim. Joel goes over to get one of the skulls from the table at the front of the room. He brings the upper teeth down onto a row of wounds in the manikin's shoulder. This is the glass-slipper moment. Do the upper canines and incisors fit into bite marks on the manikin's shoulder? And if so, do the lower teeth match a corresponding set of marks on the other side of the body?

It's a match. "Pressure and . . ." Joel positions the lower jawbone into the wounds on the manikin's backside. "Counterpressure. There's your smoking gun."

At the outset of this chapter, I mentioned a man found dead on a hiking trail with puncture marks on his neck. Investigators deemed it a cougar attack, even though there were no marks to suggest a set of matching upper and lower teeth. The wounds, it turned out, weren't made by anyone's teeth but by an ice pick. The murderer got away with the crime until twelve years on, when he bragged about it to a fellow inmate while serving time for something else.

Every so often, the opposite happens. A human is found guilty of a killing that was in fact committed by a wild animal. Most famously, there is Lindy Chamberlain, the Australian woman who screamed that she'd seen a dingo run off with her baby while the family was camping near Ayers Rock in 1980. We heard a presentation on the case from one of our instructors, predator attack specialist (and — stay tuned — survivor) Ben Beetlestone. Because the Australian investigators had no body and no dingo in custody, they could not do what we're doing today. They could not link the victim evidence to the animal evidence. Without linkage, the trial turned on assumptions (for instance, that a dingo could not or would not carry off a ten-pound baby), human error, and a media frenzy that swayed public opinion. About three years after Chamberlain was convicted, a search party looking for the remains of a rock climber found a dingo lair with remnants of the baby's clothes. Chamberlain was released and acquitted, and her conviction was overturned. The dingo really did eat her baby.

These days linkage often takes the form of a DNA match. Does DNA from the captured (or killed) suspect match DNA from hair or skin under the victim's fingernails? Does the animal's DNA match DNA from saliva on the victim? With animal attack cases, scavengers can complicate these efforts. While animal saliva near tooth marks on, say, a jacket has likely come from the attacking animal, saliva swabbed from the victim's skin could have come from an animal that fed on the corpse later.

Up in the Canadian wilderness there tend to be a lot of bears around, so good linkage is vital. Van Damme shared a story about a woman killed by a bear in her yard in Lillooet, British Columbia. His team set traps and ran DNA on two "bears of interest" before they scored a match with the third. The innocent bears were released.

It's beer o'clock (Canadian for 5:00 p.m.). Instructors are straightening tables and carrying manikins to the back of the conference room and piling them on the floor near the refreshment table. You need to straddle a corpse to get a last refill on your coffee. I waylay one of my group mates, Aaron Koss-Young, of Yukon Conservation Officer Services, for a quick overview of something that isn't covered in WHART: what people should do in an attack situation, or even just a surprise encounter. Aaron says sure. He's of the same vintage as Joel, with similar fair features and good manners.

You may have heard the ditty "If it's black, fight back. If it's brown, lie down." The idea being that brown bears, of which grizzlies are a subspecies, may lose interest in a person who appears to be dead. Right away, a problem: brown bears' fur can be black, and some black bears look brown. A more reliable way to distinguish the two is by the length and curvature of their claws, but by the time you're in a position to make that call, the knowledge will be of limited practical use. The most important thing to consider, Aaron says, is not what kind of bear you are facing, but what kind of attack. Is it predatory or is it defensive? Most bear charges are defensive. They're not really attacks, they're bluffs. You've startled the bear, or you're too close, and it would like you to back off. "It's going to come across as big and scary. Its ears are upright, not back." Aaron pauses to blow his nose. He has a miserable summer cold. "It may be swatting the ground. Huffing." Popping or clacking its jaws. (But not roaring or growling. That's mainly a movie thing.)

Aaron stuffs the Kleenex in the pocket of his fleece. "It just wants to scare the crap out of you." Grizzlies evolved in more open, less forested terrain than black bears. They often can't just disappear into the trees as a startled black bear can and typically does. So they make *you* run instead.

The recommended response to a bluff is to be as nonthreatening as you can. Back away slowly. Talk to the animal in a calm voice. You'll probably be fine — even if the bear is a sow with cubs. For all British Columbia's bears and bear encounters, and for all the hype you hear about the danger posed by protective mother bears, the province has seen only one fatal attack of that nature. (It was a grizzly. No black bear sow with cubs has ever killed a person in British Columbia.)

With a predatory attack, the survival strategy is the opposite. The rare predatory bear attack begins quietly, with focused intent. Counter to common assumption, it's more often a black bear than a grizzly. (Though with both species, predatory attacks are rare.) The bear may be following at a distance, circling around, disappearing and reappearing. If a bear starts to charge with its ears laid flat, you're the one who needs to look scary. Open your jacket to make yourself look larger. If you're in a group, get together and yell, so you look like one big, loud creature. "Try to give the message, 'I am not going to give up without a fight.' " Aaron says. "Stomp your feet, throw rocks."

The same holds true for an attacking cougar. Take inspiration from the Kansas pioneer N. C. Fancher, who in the spring of 1871 noticed a cougar eyeing him as he stood inspecting a buffalo skeleton. As recounted in *Pioneer History of Kansas,* Fancher shoved his feet inside the dead buffalo's horns, banged its femurs over his head while jumping up and down, and "bellowed desperately." The cougar, and really who wouldn't, took off.

And if the animal goes ahead and attacks anyway? "Do whatever you can to fight back," Aaron says. If it's a bear, go for the face. Aaron points in the direction of his nose, a red chapped thing. "Don't play dead." If you play dead at that point, there's a good chance you shortly won't be playing.

The worst thing you can do in any situation where a predator seems bent on attack is to turn and run. This is especially true with a carnivorous hunter like a cougar, because running (or mountain-biking) away triggers the predator-prey response. It's like a switch, and once it's flipped on, it stays on for a surprisingly long time unless a kill is made.

WHART instructor Ben Beetlestone experienced firsthand the determination and persistence of a cougar in attack mode. As a Conservation Officer in the mountainous West Kootenay region of British Columbia, he handles a fair number of predator attack calls — most involving bears and minor injuries. A few years ago, he responded to an unusual call. An emaciated cougar was skulking around a couple's property. Beetlestone shared the experience during a presentation yesterday. He told us he got out of his truck, unarmed, and went up to knock on the door, not realizing the cougar was stalking the couple *at that moment,* through their windows. "If the guy left one room and went into another," he told us, "the cougar went to that window." The windows had paw prints.

Suddenly the man slams the door. Beetlestone turns to see the cougar, five feet away, crouched, with its ears flat to its head and its tail swishing. "I'm yelling and screaming and kicking at it, all that stuff we tell the public to do. None of it is working." The cougar jumps him. He tries to choke it, but it pulls away, turns, and sinks its teeth into his work boot. He grabs a broom that's leaning against the house and hits the cougar, but it won't let go. He manages to push the broom handle down the animal's throat. Meanwhile, the couple in the house are just watching through the window. Beetlestone is holding off a cougar with a cheap tin broom, yelling, "Hey! *HEY!*"

"Finally the old guy opens the door and goes, *'What?'* And I'm like, 'I need a knife!' " The man goes to the kitchen to look for a specific knife that turns out to be in the dishwasher. Finally he finds the knife and gives it to Beetlestone, who "Bates-Motels" the cougar. (A necropsy revealed a piece of a running shoe wedged in the opening of the cougar's stomach, blocking it and starving the animal.)

The bingo game is letting out as Aaron and I collect our things and leave the conference room. One of the players, spry but slightly stooped, is making his way toward the men's room as Kevin Van Damme sets off to cross the hall with a bloody, half-naked manikin under one arm. Van Damme is an imposing figure, a purposeful strider. The bingo player halts. "Excuse me," says Van Damme, offering no explanation.

Very few cars drive the quarter-mile road from the Boomtown Casino parking lot down to the Truckee River. Today would be a diverting day to travel this road, because multiple crime scenes are cordoned off with yellow police tape. Uniformed men and women with neon-green Predator Response Team vests come and go with rifles and body bags. It's WHART field-scenario day.

My group's crime scene lies between the guardrail and the bottom of a steep, rubbly embankment. Last night we received a pretend text about an attack. Following a fight with his fiancée, we were told, a young man left the couple's Winnebago to sleep outside in his sleeping bag. At 4:00 a.m.,

35

the sheriff got a missing-persons call from the fiancée and drove out to have a look. He found the empty sleeping bag and saw a wolf, which he shot and killed. Then he turned the investigation over to a Predator Response Team. That's us.

Our first task is to secure the area, to be sure no large animals are lurking. Cougars and bears sometimes cache the bodies of their victims, burying them lightly with leaves and brush and coming back later to feed some more. This makes the "crime scene" potentially dangerous for the response team.

A young woman walks up to the man in our group who has taken the role of Incident Operations Chief. "Where's my brother?" she says. "What's going on?" It takes me a moment to realize she's a role-player. She delivers the line with no trace of agitation. More of a *Hey, what's up.* Meanwhile, at the scenario up the road, we have some N. C. Fancher–style desperate bellowing: *"YOU HAVE TO FIND HIM! HE'S A TWELVE-YEAR-OLD BOY!"* This is how it goes with these scenarios. You have one Al Pacino and everyone else is channeling C-SPAN2.

Our Ops Chief puts his hand on the sister's shoulder. "Well, we got a report there's an animal in the area."

"What kind of animal?" Like she might go back and get her binoculars. She lifts one foot to step over the police tape. "I need to be down there looking for him."

Ops Chief takes her arm gently. "Now, we don't want you to go down there and get hurt. We've got a strategic team down there, doing a diamond-shaped security sweep."

We practiced the diamond-shaped sweep earlier. Four people move along back to back to back to back, weapons ready. It's a human octopus with guns. Each person scans the quadrant in front of her (named for hours on a clock face: 12, 3, 6, and 9) and calls "Clear" if she sees no danger. Whereupon the person to her right calls "Clear." Et cetera, around and around. Not only can the surroundings be monitored in all directions, but it's safe in that no one can inadvertently point a weapon at anyone else. Should someone spot a threat, she calls it out, whereupon the people on either side move into position beside her. Now three rifles are aimed and ready, while one person watches the rear. When we practiced this earlier, Joel played the dangerous animal. I had hoped for some pantomime, maybe even a costume, but he'd just step in front of us and say, "I'm a bear."

Four of my teammates move through the

brush in the diamond formation. Aaron climbs onto a boulder to assume "lethal overwatch," his appearance of lethalness dimmed somewhat by the Kleenex wadded in the palm that supports his rifle. I'm on paperwork detail again (because "you're a writer").

"Bear, three o'clock!" It's not Joel this time. It's a lifelike bear model, one of those hard-foam target practice items used by bow-hunters. Six o'clock and twelve o'clock glide into position beside three o'clock, sliding their feet along the rough ground without looking down. They raise their weapons in unison. It's kind of balletic. It's like synchronized swimming with rifle shooting, and can we please make that an Olympic event?

On a quick count of three, pretend shots are fired at the polyethylene bear. Someone calls for tampons, and the excitement is over.

Was the wolf that the sheriff shot last night a red herring, an innocent bystander? It's our job now to figure that out. It's a wildlife whodunit.

The victim — played by one of yesterday's manikins — is shortly found down the hill from the empty sleeping bag, under a bush. A team member pretends to photograph the body, quickly, because an affable coroner, played by Joel, wants to remove it before the midday heat sets in. We'll have a chance to examine it later, at the morgue/Ponderosa Room.

Once the scene is secured, it's time for evidence collection. Items of evidence, as we know from TV police procedurals, are called exhibits. Bodies, sleeping bag, footprints, paw prints, drag marks — these are all exhibit items. Items destined for the lab are assigned numbers and put into evidence bags after they're photographed in place. A corresponding evidence flag is stuck in the ground where the item was found. My role is to note all of this — a short description of the item, its number and location — on an Exhibit Report, illegibly and probably in the wrong place.

The animal tracks in the dirt are from a bear. This is good, because we didn't learn about wolf attacks in class. (Because they almost never occur.)

The team is on hands and knees now, searching for animal hairs and blood. It's uncomfortable, hot, tedious work, but important. Much can be learned from blood at a crime scene. Round drops on the ground suggest a "gravity pattern": blood falling by its own weight from a wound.

Oblong gravity drops suggest a victim running as he dripped. A "force-related pattern" — blood ejected by the force of, say, a paw swipe or the pressure of a major artery — is elongated, with a tail like a comet. It's a spatter, not a drip.

Someone finds a trail of drips. Joel tells us to look closely at their size. When drips of blood grow smaller as the trail progresses, they're probably not coming from a wound. They might be dripping from the animal's fur, or a murderer's blade. If the size of the drips remains constant — a "replenishing trail" — they are likely coming from an "active bleeder." A smear of blood is a "contact pattern," perhaps a place where the victim fell or placed a bloody hand.

When we're sure we've found everything there is to be found, Joel reaches down and flips over a leaf, revealing a tiny drop of blood on the underside. We missed this. We missed a lot — blood on rocks, plants, on the ground. "Splatter pattern," someone says knowingly.

Joel nods, but adds quietly, "S*patter,* not *splatter.*"

Together the blood and the marks in the dirt tell the narrative of the attack. Drips and blood on the sleeping bag from the initial bites. A drag mark and replenishing drips as the man is pulled from the sleeping bag into the brush. Scuffle marks and blood in the dirt as the man tries to escape, and then a spatter pattern on the plants and rocks, perhaps caused by the bear shaking the man to stop the struggle. Had the body lain dead for any length of time, the chemicals of its decay would have left a final piece of evidence, a stain or area of blackened vegetation, called a "decomposition island." No pretty beaches there.

Our victim's injuries, Joel tells us, have been recreated on one of the manikins. It's not here at the scene, but we'll examine it in class tomorrow morning when we try to establish linkage.

And that brings us to beer o'clock again. Joel collects the props and the evidence flags and the polyethylene bear, and we all troop back along the road and up to our hotel rooms to change. By the time I come back downstairs, my group has gathered at a small sports bar behind the blackjack tables. They're intent on hockey, Oilers versus the Toronto Maple Leafs.

"Hey," I try. "Shouldn't that be Toronto Maple *Leaves*?" I can't compete with hockey, so I go for a walk. I end up at a Cabela's outfitters. I don't hunt, but I enjoy the taxidermy. This outlet has an outstanding mountain diorama and a musk ox on top of the dressing rooms. Also a Gun Library, which, I discover, contains used guns, not books.

The man behind the counter waits for me to say something. I ask about getting a library card. "You can't borrow these guns," he says. "They're for sale."

"Then it's not much of a library, is it?" Seems like I should probably call it a night.

The manikin from our crime scene comes with some extras. Joel has just emptied onto a tabletop a bag of realistic moulage bear stomach contents: an ear and an eye and a strip of scalp with part of a mohawk haircut. These are passed around among our group. It's early in the morning for such things. Doughnuts sit untouched.

The stomach contents are a match for what's missing on our manikin's head, suggesting that indeed the bear, not the wolf, was behind the attack. The mohawk seems like a fanciful touch, but turns out not to be. Joel reveals that our scenario from yesterday was based on an actual attack — real bear, real man, real mohawk. Joel investigated this case in 2015. All the WHART manikins, in fact, represent not just real wounds but real attack victims.

Joel brought along photographs from the actual attack scene. One shows the victim's backside. The largest wound, a raw, gaping, messy chomp, is to the buttocks. The man had been sleeping in one-piece long johns, and the flap, Joel says, must have opened while the bear was dragging him. "So that's why there's feeding right there." After a moment, Joel adds, "You know the one with the bear paw prints on it? On the butt flap?" This is apparently a common item in Canada, because several of my group mates nod. "That's what he was wearing."

There's a clean set of bite marks on the manikin's shoulder. From the position of the upper and lower canine marks there, we can tell that the man had been sleeping on his back. The bear, Joel surmises, came upon the sleeping figure, maybe licked the salts from his skin. The man woke up and probably made some noise. "So the bear figures, *Well, I either finish this or I run away.* He chose to finish it."

Meanwhile, what was inside the stomach of our other suspect, the wolf shot by the sheriff when he arrived on the scene? Gum wrappers and tinfoil. No human tissue or clothing. Case closed. No DNA analysis was needed.

Once the forensics is completed and the

perpetrator known, what happens next? If this bear hadn't been shot near the scene of the attack, what would have been its fate? Kevin Van Damme talked about this after a lecture. Prison isn't an option. Canadian zoos won't take bears older than three months, because they tend to pace and because zoos generally have enough bears. Capital punishment is what happens. "If a bear treats a person as food, it will do it again," Van Damme said. "I have spent twenty-six years as a predator attack specialist. I know some of you disagree with me, but if it hurts a person, it's going to die."

As any criminologist can tell you, prevention is better than punishment. The safest thing for both species is to keep them apart. Don't let bears learn to associate humans with easy meals. Require that people in bear country secure their garbage. Tell them to stop feeding birds and leaving dogfood on the porch. The man in the long johns lived in the woods, where there was no garbage pickup. Trash likely piled up outside the trailer. The tinfoil and gum wrappers in the wolf's stomach suggest that this was a place wild animals had become comfortable scavenging for scraps. Garbage is a killer.

2
BREAKING AND ENTERING
AND EATING:
HOW DO YOU HANDLE
A HUNGRY BEAR?

Stewart Breck is a tall, narrow plank of a man. His arms don't stray far from his sides as he walks, and he carries no backpack or bag to break up the long vertical plane he occupies in space. You notice this when you walk behind him, which I've been doing a lot of because his stride covers several city blocks. Though personable, his demeanor, too, tilts to restraint. Over the course of the day I've spent with him, he has not raised his voice or gesticulated memorably or used a bleepable word. He's composed, considerate, reasonable. I'm telling you this so you'll understand how I was a little shaken when Stewart Breck, a moment ago, went, "Are you *FUCKING KIDDING ME?*" and his arms shot out from his sides, where they remain, palms up, the universal gesture of exasperation.

Because I am, again, lagging behind, I don't at first see what Breck sees. Now I do: two fat trash bags ripped open, with food scraps spilling out onto the pavement. It is 3:30 a.m., bear time in the back alleys of compact, restaurant-dense downtown Aspen, Colorado. The sound of Breck's approaching SUV must have scared off a bear mid-scavenge. Compost and garbage are known in the parlance of human-bear conflict as "attractants." Aspen municipal code requires both to be secured in bear-resistant containers.

"Give me a break." Quieter now, the hands back at his sides. "We spent hundreds of thousands of dollars on this." *This* equals: multiyear, multicity research into how best to get people in the midst of bear country to properly lock up attractants, and how much difference it makes when they do. The work was funded by Colorado Parks and Wildlife (CPW), who get the calls when bears damage property while looting unsecured human food; Colorado State University, where Breck teaches a course in human-wildlife conflict; and Breck's employer, the National Wildlife Research Center (NWRC), headquartered in Fort Collins, Colorado.

NWRC is the research arm of Wildlife Services, which is part of the United States Department of Agriculture (USDA). The "services" are provided mainly to ranchers and farmers who are having problems with wildlife cutting into their livelihood, and often they take the form of killing that wildlife. Breck was hired by NWRC to research nonlethal alternatives. His job gives him lots of opportunity to deploy his admirable composure. There are old-schoolers at Wildlife Services who hate him for rocking the boat, and there are animal welfare activists who hate him for not rocking it hard enough. I like him because he's trying to stand on the impossible middle ground.

What the garbage studies showed is that reinforced, locking bear-resistant containers make a solid difference — provided people take the time to latch them properly. In an area where 80 percent of the containers were used as they're meant to be, there were 45 human-bear conflicts over the course of the study. A similar area with only 10 percent compliance had 272 conflicts. What this says is that containers aren't enough. You also need laws requiring people to use them, and fines for people who ignore those laws. Aspen has all of this, but there has been a reluctance to follow through with the fines. Especially here, downtown. Breck has been told that in the intervening years the situation has improved.

Just now, it's not seeming that way. Coming down the alley in that unhurried, endearingly pigeon-toed way is a full-grown black bear. Breck and I are standing near his vehicle, which is parked twenty feet back from the mess. The bear nears the garbage, which has been its focus until this moment, and then it looks over at us. It clacks its jaws, an indication that it's uneasy. For here are two staring humans, one with some good height to him, at a time of night when humans are rarely about. On the other hand: kitchen scraps from Campo de Fiori! The bear considers the situation a moment longer, then lowers its head to eat.

Because there's a lot of eating to be done. It's early fall, the time of year when black bears eat with purpose and abandon, to construct the fat they will live off in their dens over the winter.* A hyperphagic black

* You may be wondering: When you live off your own fat, do you need to use the toilet? If you are a bear, you do not. Hibernating bears reabsorb their urine and form a "fecal plug." Cubs, on the other hand, let it go inside the den. Not a problem, because the mother bear eats it — partly as cleanup, but mostly as food. She is nursing, after all. While hibernating. Black bear hibernation isn't the same as sleep. They're just sort of slowed down

48

bear doubles or even triples its daily calorie count, taking in as much as 20,000 calories. As omnivores, bears happily eat a variety of foods; during hyperphagia, what they are drawn to most powerfully is a concentrated source. They want to take in lots of calories without having to burn lots of calories wandering around looking for calories. The mountains around Aspen have always supplied that: acorn-dropping oak brush, fruiting serviceberry and chokecherry trees, the outrageous fecundity of crabapple trees. Come the 1950s and '60s, the skiers began to move in. Bears looked up from their nuts and berries and went, *Hurunh? Birdseed hanging on a tree? Bag of kibble sitting on a deck? Yes, please.* Soon they ventured into town, following the humans, because the humans provide. The alleys behind Aspen's

---

and out of it. Surreally, black bear sows give birth halfway through their hibernation. They deliver a couple of cubs, snack on the placenta, then go back into hibernation, nursing and tending their cubs in a state of semi-alertness until spring. According to a scientist who has taken blood samples from hibernating black bears, they do not have sleep breath and their dens don't stink. They smell like roots and earth and that is all.

multitudinous restaurants are concentrated-food-source nirvana.

Breck nudges me. Another bear is coming down the alley, this one darker and slightly smaller. The lighter, dominant bear turns its attention to the newcomer and makes a low, rumbly sound. *You may have those hearts of romaine and that spinach gnocchi, but do not come near my grilled sustainable Skuna Bay salmon.*

Breck raises his phone to take a picture, which surprises me. This is a man who uses the word *routine* to describe the act of hand-darting a hibernating black bear to replace its tracking collar. It turns out he's not photographing bears. He's photographing irony. "Look at the lid." He aims his flashlight at the wheeled compost cart lying, open, on its side. The molded-plastic lid features a bear face, and inches from this decorative bear face is the face of the actual bear now enjoying the contents of this certified bear-resistant container that has failed to resist it.

"They jump on them," Breck says, "and they pop open."

Or the locking mechanism may be broken. This was the case with another of this same model of compost cart farther along the alley, which we saw earlier in the day. Breck walked up and lifted the lid on fifty reeking bananas. "Be sure to latch," chided a sticker. "A bear's life depends on it." In the next alley over, Breck led me to an uncovered vat of used cooking grease. It was as tall and big around as a drinking fountain, and bears sometimes use it like one. Breck has seen paw prints in grease leading away down the alley.

Chapter 12.08 of Aspen's solid-waste code, entitled "Wildlife Protection," was modeled on that of the neighboring ski and mountain-bike resort village Snowmass. There the similarity fizzles out. Snowmass Animal Services/Traffic Control consists of Tina White and Lauren Martenson, and they are *on it*. "We ticket everyone," White told me when we met yesterday. She recently put together a slide presentation in Spanish for restaurant kitchen staff, many of whom hadn't realized what happens to bears that start raiding dumpsters when people neglect to lock them. Her efforts have been working. It's been several years since a bear causing problems in Snowmass was, as White put it, "pulled out of the mix." At the time of my visit, Aspen was up to nine for the year. Then again, Aspen is three times as populous, with four times the number of restaurants.

Aspen's garbage violations are handled by community response officers, five in all. Breck and I met with their representative, Charlie Martin, yesterday morning in a conference room at the Aspen Police Department. Charlie wore a black and yellow uniform, and a pair of socks on which rainbows alternated with unicorns. "It's not Friday and I wasn't on bike patrol," he said, mysteriously, when I commented on them. Charlie listed some of the things his team was already struggling to keep up with when bear-related garbage infractions were added to the list: traffic violations, barking dogs, idling construction vehicles, 911 calls, rabid bats, lost and found, sidewalk snow, jump starts, vehicle lockouts, community picnics, and removing dead deer from roadways.

Charlie was a trifle defensive about the alley situation. "We've passed out almost ten thousand dollars in tickets this year." The fines for leaving garbage or compost unsecured range from $250 to $1,000. Breck and I could have matched the year's total in one day. Except that, as Charlie pointed out, the fines wouldn't stick. "You've got multiple parties sharing one dumpster," Charlie said, referring to containers in both condo developments and restaurant back alleys. "You write someone a ticket, and they'll say,

'It was someone else. We left at ten p.m., and we locked it. Prove to me that it wasn't locked when we left.' ”

Aspen's waste management companies are, by law, required to assign a number to each compost and trash container, and to keep a database that links these numbers to the person or company responsible for keeping the contents secured and paying a fine if they are not. Aspen contracts with five of these companies, and none appears to have set up such a system. (Snowmass does its own pickup. Also, Tina White will happily climb into a dumpster and rummage through a trash bag for mail with a name and address. She has heard people refer to her and Lauren as "the bear bitches.")

You read about this kind of thing over and over in communities that have tried to switch to bear-resistant containers. Generally speaking, waste management companies are fiercely concerned with their bottom line, and not so fiercely concerned with the welfare of bears. The containers need to fit the lifts on the trucks, which means that on top of the expense of the bins, there will be the expense of new trucks or retrofitted trucks, and either way it's money the companies would prefer not to spend. And the people who respond to the bear calls are not the people who draft the ordinances or the people who run the garbage companies. It's a stinking mess.

While wandering the alleys this afternoon, Breck peered over the lip of a dumpster marked CARDBOARD ONLY. At the bottom were French fries, an olive, and some squeezed lemon halves. As it's written, city code doesn't require recycling dumpsters to be bear-resistant or locked or even covered, and people often toss in bags of trash. On the residential side, problems arise when homeowners rent out their property and the vacationers either aren't told about, or don't remember or don't care about, the garbage laws.

Charlie agrees with Breck that Aspen needs an overhaul. It needs to replace the busted bear-resistant carts for the downtown compost and trash. It needs to resolve the shared dumpster loophole. Most importantly, it needs to hire enough staff to stay on top of things.

That would not, Breck added, be a heavy lift for Aspen. The county is home to about as many billionaires as bears. The Koch brothers have property here. The Bezos parents. The Lauder siblings. There is oil money, hedge fund money, cosmetics money, tech money, lingerie money, tinfoil money, chewing gum money. Breck believes that may be part of why the Aspen enforcement effort stumbled, that the city's council members kowtow to its alpha residents.

Of course, the billionaires don't manage the restaurants. That part may be Charlie's fault. "I've got to live in this town too," he said to us at one point. "And I'd like to go out to a restaurant and eat. I just gave them a thousand-dollar ticket, and I'm gonna go in their restaurant?" Aspen needs bear bitches.

The lighter-colored bear is working a crab leg, while its colleague noses through cabbage leaves. "What have these bears just learned?" Breck is saying. *"I can eat garbage with people standing and watching me and nothing bad happens."* When Breck first joined the National Wildlife Research Center, he did some human-bear conflict research in Yosemite National Park. In the park's early days, he says, staff would set up bleachers and lights around the garbage dump and charge visitors for the show: twenty or more black bears gorging and pushing each other around.

Right now we're the people in the bleachers. We've just given these two a little less reason to worry about humans. As a result, they may start coming into the alley earlier or standing their ground longer. Odds are they'll end up like the bear that dined out at the dumpster behind Steakhouse No. 316. One night not long ago, the restaurant's manager, Roy, came out to roust the animal. Because the dumpster was set in an alcove, the bear's escape was blocked on three sides. On the fourth side was Roy. With only one way out, the bear lunged and, quoting Charlie, "bit Roy in the ass." According to University of Calgary professor emeritus and bear attack researcher Stephen Herrero, 90 percent of black bears that injure humans are bears that have habituated to them — that is, accustomed to their presence and lost their fear — and developed a taste for their foods.

Based on a description of the bear that Roy provided, the animal was found, trapped, and, because it had injured someone, put down. (What the description said beyond "dark hair" and "heavyset" I can't imagine; however, DNA from saliva on Roy's pants was a match with the bear's.)

Roy and his staff could have been more careful about keeping the dumpster locked and that, too, bit him in the ass. Townspeople picketed the steakhouse following the bear's death. People don't want bears destroyed because of other people's neglect. If anything, they want them hazed or relocated — the two nonlethal approaches you hear about most with "conflict bears." (There's also electric fencing, but the prison-camp look doesn't play well in residential areas.)

Hazing refers to the practice of supplying a frightening or painful experience such that the animal associates the unpleasantness with the location or the behavior underway when it began, and then avoids such in the future. In the case of these two bears, you'd need to station someone here in the alley during the wee hours with an implement of less lethal unpleasantness,* most likely a

* Taser International briefly sold a wildlife taser, the X3W, that some thought held promise as a hazing tool. The devices were purchased mainly to be used as they are on humans, to gain control in a threatening situation without necessitating the firing of a lethal weapon. The item sold poorly, a company representative told me, because it was costly and because it only worked on very tall mammals — a moose, or a bear on its hind legs — and at a distance of less than 25 feet. (Otherwise one of its two probes, the downwardly aimed one, would hit the ground.) The impetus for the X3W was an agitated moose whose calves were

gun that shoots rubber ammunition or bean bags. If you are a law enforcement ignoramus like me, you may be picturing the colorful handsewn item tossed at holes from a distance or juggled by clowns. These bean bags are smaller, about the size of a walnut. They don't penetrate skin, or hide, but they smart.

"Hazing is never going to solve this," says Breck. The bigger bear rips deeper into his

---

stuck in an open house foundation. The moose had chased Larry Lewis, of the Alaska Department of Fish and Game, and a state trooper three times around the patrol car, when the trooper pulled out his Taser. The moose was stunned, recovered, and ran off, allowing Lewis to safely extract the calves. Impressed, he contacted the Taser people and worked with them to design a wildlife version that was safety-tested at the Kenai Moose Research Center ("a world leader in moose science"). Tasing was found to be less stressful to the animal than tranquilization, and safer in that there's no risk of killing it by overdose. (Darts are filled according to an estimate of the animal's weight.) So it seemed like a promising alternative when a situation develops quickly and lives are at stake, or when a moose has "a chicken feeder stuck on its head" — the example Lewis provided for *Alaska Fish & Wildlife News.*

garbage bag. "There's too much to be gained." How well hazing works depends on the push and pull of risk and benefit. These bears have learned that a visit to this alley is likely to offer a caloric windfall. Weighed against those calories, the risk of another smack to the flank would be a risk worth taking. "And there's too much other stuff nearby," says Breck. "If you were to haze these bears right here, they'd just go over to the next alley."

When hazing does work, it generally does not do so for long. In 2004, a team of Nevada wildlife biologists assessed the effectiveness of hazing black bears in an urban locale. One group of bears was hazed with rubber bullets, pepper spray, and loud noises, and another group got all that plus a barking Karelian bear dog to run it off. A control group was not hazed. In terms of how much time passed before the bears returned, neither group stayed away significantly longer than the group that hadn't been hazed. All but five bears out of the sixty-two followed in the study showed up again eventually, and 70 percent were back in fewer than forty days.

Breck spent many a late night trying to haze bears during an epidemic of car break-ins at Yosemite campgrounds. Between 2001 and 2007, eleven hundred automobiles were broken into by bears. (Minivans were hit most often. While it's possible some structural weakness contributed, Breck believes it had more to do with what minivans typically hold: kids, lots of them, spilling juice, dropping crumbs, grinding chips into the footwells. He guesses the bears were keying in to the smells of this "micro-trash.") Hazing efforts proved futile. "Once they learn what's inside . . . forget it." The bears quickly came to recognize the sound of Breck's truck. They'd take off when they heard it coming, and go back when they heard it drive away.

It turned out that fewer than five bears — sows and their cubs — were behind the break-ins. This is typical. From the start of the year to the September of my visit, bears in Snowmass have broken into houses through unlocked doors or windows sixty times. Wildlife-camera images have implicated just four bears. A bear research scientist with the Minnesota Department of Natural Resources, Dave Garshelis, told me about a call he got from a National Guard camp where bears were raiding pallets of military rations called MREs, which bears apparently enjoy more than soldiers do. He was told that around a hundred bears were raiding the supplies. "The guy said, 'I'll bring you to this place where you look across to this ridge that is completely pockmarked with bear dens.' I was like, 'This sounds cool.'" The "dens" turned out to be natural landscape features, and "a hundred bears" in fact was three.

Great, so just trap the few brigands and deliver them deep into the forest, and your troubles are over, no? Say hello to the disappointing reality of translocation. Adult black bears rarely stay put where they're released. They have made their way home in journeys as lengthy as 142 miles — in one case including a 6-mile ocean swim. It is a remarkable achievement given that, unlike migrating birds, they can't rely on internal magneto-gadgetry to help them navigate. Whether they are picking up sensory cues — the smell of the ocean, say, or the sound of an airport — or just trying out different directions until something feels familiar, is not known, but they are motivated and they are good at it.

In a 2014 study, sixty-six conflict bears were radio-collared and translocated by Colorado Parks and Wildlife. Thirty-three percent of the adults made their way back to the spot where they'd been captured, and none of the subadult bears did. Those read as fairly optimistic statistics; however, if you define success not as the failure to return but rather as surviving a year in the new home, the picture is less rosy. Translocated bears often wander into a new town close to where they've been released and start getting into the same kind of trouble. More than 40 percent of translocated bears in Yellowstone National Park and 66 percent in Montana were involved in another "nuisance event" within two years. Yosemite rangers tried translocating the bears that were breaking into cars, moving them to the other side of the park. The result: car breakins on the other side of the park.

Another factor is at play in the decision. Were a person to be seriously harmed by a translocated animal in its new location, the agency that brought it there could be held partly liable. The Arizona Game and Fish Department settled out of court for $4.5 million after a bear they'd translocated mauled a young girl at a campsite.

Dave Garshelis has worked with humans and bears for almost forty years. I asked him, by phone, how he felt about translocation. "People think this is a kind thing to do, but I'm not sure it is all that kind," he said. Often it's sows with cubs that get into trouble, because they need the most food.

"Here she is living in her home range, teaching her cubs where the foods are. Now all of a sudden you plop her down somewhere else that she's completely unfamiliar with. With a whole bunch of other bears which she's competing with for food. You're injecting them into a social system they're not familiar with." When bear biologists from the state of Washington surveyed forty-eight U.S. wildlife agencies, 75 percent said they sometimes translocate problem bears, but only 15 percent believed it was an effective way to resolve the problem. It's more often done in high-profile cases, when media attention has put the animal and the agency in the spotlight. Generally speaking, translocation is a better tool for managing the public than it is for managing bears.

The most promising candidates are young bears translocated early in their "criminal" careers. This is partly because yearlings are less inclined, or less able, to find their way back, but mainly because dumpster diving is a gateway crime. Next comes breaking and entering, burglary, home invasion. As garbage-eaters become habituated to humans, as they start to associate them with jackpots of food, the risk-benefit ratio shifts. Less perceived risk, dependable benefits. Why stop with the metal boxes in the restaurant back alleys? Why not get inside the big boxes in the hills with the enticing cooking smells? Since the end of hibernation, in April, Colorado Parks and Wildlife has had 421 calls about Pitkin County bears damaging property while going after people's food. Most of these calls go to District Wildlife Manager Kurtis Tesch, whom Breck and I are meeting up with tomorrow.

The darker bear, perhaps weary of being harried by its dominant associate, has snagged a bag and run up a short set of steps. We follow it up and around a corner, to the upper level of a swank mini-mall. Ordinarily, I would take delight in the optical non sequitur of a bear standing in front of a Louis Vuitton boutique. This poor goober with the burrata on its snout, innocent and utterly unaware of its likely fate, makes me want to cry.

Kurtis Tesch has bear stories, but maybe not the kind you expect. The things that stay with him are not the displays of strength or violence but rather the intelligence and occasional unexpected lightness of touch. The bear that unwrapped the foil on a Hershey's Kiss. A bear that stood up, grasped a door on either side and pulled it from its frame, then carefully leaned it up against the house.

"They'll reach in and take things out of the fridge, like eggs, and set them aside without breaking any." We are on the way to the scene of a break-in high on a ridge road, Kurtis and Breck and myself, careening around switchbacks in Kurtis's cluttered, thrumming CPW truck. An egg wouldn't last long in here.

Black bears are keeping Kurtis unusually busy this year. This was unexpected, because the spring was wet; human-bear conflicts are typically thought to intensify with drought, not with plentiful rain. But the year before was very dry, and Kurtis says he's heard that drought spurs some plants to produce an excess of reproductive material, or "mast" — fruit, seeds, berries, acorns — and then less of it the following year. "They're trying to spread their seed, thinking that they're about to die off. And then when a wet year comes, they're more concerned about growing." I don't know if that's what has happened here, but I like this worldview of trees that worry and prioritize and plan for their demise.

From the back seat, Breck volunteers that the general trend toward warmer temperatures also contributes, by shortening the length of the bears' hibernation. In a 2017 study, he and six CPW biologists radio-collared 51 adult black bears and monitored the timing and duration of their hibernation, along with environmental factors. For every 1.8-degree Fahrenheit increase in temperature, hibernation shortened by about a week. Based on current climate change projections, black bears of the year 2050 will be hibernating 15 to 40 days less than they are now. That's 15 to 40 more days out on the landscape looking for food. Add "more bear break-ins" to the list of possible consequences of climate change.

Food supply also affects hibernation. In a year of plentiful food, bears hibernate for shorter periods. For a bear that starts relying on human-sourced foods, every year is a plentiful year. Breck found that bears that foraged mostly urban areas hibernated a full month less than bears that foraged the natural landscape. Another concerning consequence of plentiful food is that reproduction rates rise. Black bear sows have a reproductive option called delayed implantation. Fertilized eggs become clusters of cells, called blastocysts, that loiter in the uterus over the summer. Whether they implant in it come fall — and how many of them do so — depends on the mother's health and how well she's been eating.

We've arrived at the driveway of our destination. From here, the house looks to be of average size. Turns out this is because most of what we're looking at is garage. The house pours down the mountainside two, three, I don't know how many stories. Breck steps down from the truck and walks to the edge of the blacktop. I assume he's marveling at the view, but as I walk over I hear him calling out the names of bushes and trees growing wild around the house, the ones black bears feed on: serviceberry, chokecherry, oak.

"Yup," says Kurtis. "This is some of the best bear habitat in Colorado. We moved into their habitat. You know?" Kurtis wears reflective orange-tinted sunglasses that stay on his face the whole time we're with him. He's light-haired and fit, with a good jawline, and that's as far as I can take you.

The owners of the house have been out of town. The housekeeper, Carmen, discovered the break-in and called the police, who in turn called Kurtis. Carmen lets us in and takes us downstairs to the entry point: a floor-to-ceiling window in a bedroom with its own deck. She says it was locked, but bears can wedge their claws into any small gap in a window frame and pry the unit out. An interior screen window lies on the carpet. The wall-to-wall is white, but the bear left no tracks. Nor, Carmen says, did it knock anything over on its way upstairs to the refrigerator. You get the sense that if there'd been a mop handy, it would have cleaned up the kitchen floor.

This bear reminds Breck of one that was breaking into Aspen homes back when his study was underway here. They called him Fat Albert. "He was just kinda laid-back. He'd gently open a door of a cabin, go in, eat some food, and leave. People would go, 'Wow, he didn't destroy my place at all.'" That's why he was fat, and that's why he was alive. There's more tolerance for a bear like that. An aggressive bear that trashes the place or otherwise makes homeowners feel violated and in danger is very quickly going to be, to use Breck's word, whacked. The upside, if it can be said there is one, is that natural selection favors the Fat Alberts. Aggressive bears are likely to be put down before they have much opportunity to pass on their genes.

With a growing percentage of Fat Alberts, will coexistence eventually become a possibility? Or even a policy? Could we live with bears in the backyard the way we live with raccoons and skunks? I posed this question to Mario Klip, a bear specialist with the California Department of Fish and Wildlife (CDFW) in the Lake Tahoe region. Many people in his area already do, he said. Say a couple of homeowners find a bear under the deck. Rather than call Fish and Wildlife, they may call the Bear League, a local advocacy group. "They'll send someone out to crawl under and poke it with a stick and get it to run off, and then help board up the space for you."

Klip practices coexistence with the BEAR League. "They are," he points out, "filling a vacuum." More and more people want nonlethal options for bears that trespass or break into houses. And not just Californians. Dave Garshelis works in rural northern Minnesota, where most people have guns and are allowed — encouraged, even — to solve their bear problems themselves. "I've been here thirty-six years," Garshelis told me. "I can sense a sea change in attitudes to bears."

What would happen if wildlife managers did nothing, if they stopped destroying the recidivist bears? The fear is this: Those bears' cubs will learn to break into homes, and ditto those cubs' cubs. As break-ins escalate, tolerance erodes. As Garshelis put it, "It's hard to be tolerant when there's a bear in your kitchen."

Back upstairs, Carmen describes the scene as she found it. The bear appears to have gone straight to the refrigerator. It opened the door, pulled out and scarfed a tub of cottage cheese, broke a bottle of maple syrup and a jar of honey and lapped those up, and then moved on to a pint of Häagen-Dazs in the freezer. (Pitkin County bears consistently prefer premium brands. "They will not touch Western Family ice cream," Tina White reports.)

Behind us a set of French doors leads to another deck. Carmen found these doors open and assumes this is where the intruder left the house. French door handles, locked or unlocked, are so easy for black bears to open that they're known as "bear handles" and are prohibited by local building code. But people like them, and do-it-yourselfers either don't know or don't care about the finer points of building code, and Kurtis sees them everywhere. Hollow doorknobs are likewise prohibited; bears crush and grip them in their teeth and easily turn them. (Some businesses make things even easier. Automatic doors open for bears, too.)

Kurtis thinks we may be looking at the work of two different bears. The first one entered and exited through the downstairs bedroom window, and a different bear came up to the French doors on the kitchen deck and smelled or saw the aftermath of the first pillage. His reasoning is based on the position of the doors as Carmen found them: opened inward. It would be unusual, he says, for a bear to pull a door inward in order to pass through. It's also possible the same bear returned to the scene a second time. Kurtis says they often come back at least once.

Like human burglars, bears typically break in when the homeowners are away. Given the large percentage of Aspen properties that are let as vacation rentals part of the year, empty homes are easy for bears to find. With bolder bears, burglary may escalate to home invasion. Often the bear comes in while people are asleep, especially, Kurtis says, when it's hot and someone has left the windows open. Or a sliding door is left unlocked. Sometimes the residents are not asleep. "We've had people eating dinner at their table and the bear walks in, grabs some food off the table, and runs back outside. We've had bears ripping doors or windows out while people are in there, hiding in their bedroom or bathroom."

Kurtis gives Carmen his card and tells her to have the owners call him if they want a live trap set. She doesn't ask him what would happen to a bear that ends up in the trap. Colorado Parks and Wildlife, like many state wildlife agencies, has a two-strike policy. If Kurtis gets a call about a bear nosing around someone's trash or hanging out in a back yard, say, he will attempt to trap it, and if he succeeds, he'll ear-tag it and take it into the woods and release it and hope it doesn't come back. (A trap is left in place no more than three days, to lower the odds of trapping the wrong bear.) Often the trap stays empty. "We're not catching them like we used to," Kurtis confided later. "I don't know if they've just gotten smarter, or what the deal is."

The bear that broke in here would not be granted a second strike. Because it's breaking into locked windows, and will likely continue to do so — and, if it's a sow, teach its cubs to do so — the agency considers it a threat to public safety. Kurtis says people often decline to report break-ins, because they know their call may set in motion a death sentence for the bear. The black bear is a ridiculously lovable species. There's a reason kids have teddy bears, not teddy goats or teddy eels.

"So what would happen if you were to trap this bear?" We're climbing back into the truck now, to head back to town. I notice a lint roller in the door caddy, as if sometimes bears sat up front in the cab.

"When you do trap the right one and take it out," Kurtis says, "you notice a slight decrease in break-ins in the area. For a short period of time. And eventually another bear comes in and takes over. So."

"It's a temporary solution," says Breck. "You're just mowing the grass."

That wasn't exactly what I was asking about. I was more asking about the "taking out." I'm going to have to be more direct. "And it can't be fun to have to put down a bear." All these euphemisms.* Taking out. Putting down. Are we killing an animal or unloading a truck?

"No, it's not," Kurtis says flatly. "Last

* I understand the desire on the part of those whose job sometimes requires them to kill an animal to avoid that verb. *Kill* has a taint of murder. The sheer number of euphemisms suggests a long-running struggle to find something right. I collected them for a while: *cull, take, dispatch, remove, lethally remove.* As a word person, I balk at *euthanize,* which implies the relief of suffering, and *harvest,* which makes animals sound like corn. I heard one person say *"use lethal force on,"* which seems better suited to SWAT operations and Gary Busey movies.

week I had to put down a sow and a cub." The pair were repeatedly breaking into houses. "And that is not fun. At all." We drive along in a grim quiet broken intermittently by the walkie-talkie clearing its throat.

"In that instance," Kurtis adds, "I was struggling on how to do it exactly. I didn't want to put the cub down and have the mom watch that. I didn't want to put the mom down and have the baby see that. I ended up darting the baby so it went to sleep. I put the mom down, and then while the baby was asleep, I put it down. That way neither one had to see the other. So."

When Kurtis says "so" it is shorthand for the many frustrations the job holds. Uncaring homeowners who don't bother to follow the laws. Who then blame and hound him when a bear crosses a line and is destroyed. Government agencies that would rather pass the buck than spend a buck.

I try to imagine how I would feel if I lived in the house we just left and I had seen how effortless it was for a bear to get in. I ask Kurtis how people usually react. "Some are terrified," he answers. "Some are nonchalant." So far, no one in the area has been killed during one of these break-ins. Black bears, by and large, are not aggressive animals. Still, I'm surprised that what sometimes happens when human burglars break into homes hasn't happened here: homeowner or homeowner's dog surprises burglar, homeowner and/or dog goes after burglar, burglar panics and kills homeowner.

"Oh, it's coming," Kurtis says. Black bears may be no more aggressive than raccoons, but they're a lot larger.

What if we accepted that risk? What if we chose to live not only with the occasional bear in the kitchen but with the likelihood that someone at some point will be killed by one of those bears? Planes are allowed to operate even though every now and then they crash and people die. One difference is that with airlines, sales revenue covers the expense of lawsuits and insurance. When a bear harms or kills a person, the state wildlife agency may be held liable, and bears, unlike planes, aren't generating the revenue it would take to cover the costs. There have been a couple of lawsuits recently, one in Utah and another in Arizona, with large payouts to the families of the victims. The agencies had been aware of the bear's presence in the area but had opted to monitor the situation rather than set a trap.

Breck rolls down his window. "So that's going to be your limiting factor on that idea."

■ ■ ■ ■

Away from the back alleys, Aspen is pictur-esque and pristine. There are almost as many window boxes as windows, and although we're closing in on October, I have yet to see anything in them that's starting to wither or turn brown. It's like this town has so much money and power that even the laws of nature shrug and surrender. Flowers bloom in fall, and women's hair goes ash-blond as they age.

Where I see prettiness, Breck sees at-tractants. "These right here?" Breck points over our heads, to one of the small trees that line the pedestrian walkway where we've been questing for an affordable lunch. "Crabapples. The city *planted crabapple trees.*" People enjoy the profusion of pink spring blossoms. That then turn into bite-sized apples that bears mouth straight off the branch, like cartoon emperors with their clusters of grapes. Black bears show up mid-day in downtown Aspen regularly enough that the city passed a law making it a ticket-able offense to ignore the CPW officer standing guard and go right up to a bear and snap a selfie. Kurtis Tesch's predeces-sor tried, unsuccessfully, to convince the city council to replace the crabapple trees. When I got home I stumbled onto an online Aspen Arbor Guide to residential tree selection and planting. Among the recommended species: crabapple, oak, chokecherry, and serviceberry trees. I hesitated to tell Breck. I thought his head might explode.

We locate a moderately priced restaurant that is not one of the eighteen fined for having unsecured garbage and publicly shamed in this week's *Aspen Times*. I dig out a list of questions, questions that basically boil down to: What is happening here, and is there an answer? I bring up something Kurtis Tesch said when we were driving back to town. He shared a theory about how the ballot measure that banned Colorado's spring bear hunt (because it orphaned cubs) had caused the rise in the number of bear conflicts. Breck says he often hears this argument. "There's a sentiment carried forward by a lot of hunting communities and parks and wildlife people that one way out of this is to hunt our way out of it. But there's no good science to say that lowering the bear population will lower the number of conflicts."

For one thing, he says, there's a mismatch between where the hunters go and where the conflicts are. "Hunting quotas are set according to game management units." I don't catch all the details because I'm distracted by some intense celebrity name-dropping at a large table next to ours. I hear: "This game unit has Aspen, Snowmass, Carbondale —" ". . . Reese Witherspoon —" "They'll say, 'Okay, within this unit, harvest x number —' " "So then Reese . . ."

Hunting does alter the behavior of the hunted somewhat, Breck allows, in that it perpetuates fear and avoidance of humans. But Colorado still has a fall bear hunt, so he doesn't buy less hunting as a reason for the rise in bear conflicts in this case.

It's worth mentioning that the money for Kurtis's salary — like most items in state fish and wildlife agency budgets — comes partly from hunting and fishing license fees and taxes on equipment. "I'm not here to criticize that model," Breck says, "but you have to be aware that that's underlying all this stuff."

I am. Somewhat uncomfortably so. In the course of researching this book, I met a lot of good, intelligent people at these agencies, professionals who saw their job as protecting people and animals both. But because of the financial model, it can be hard to set aside the nagging sense that institutional priorities are at play. The money's coming from hunters, to a large degree — and that makes it hard for agencies to win the trust of everyone else. (And creates perplexing mottoes like "Support Nevada's Wildlife . . . Buy a Hunting and Fishing License.")

Breck shakes open his napkin. There's legislation making its way through Congress that would add a billion-plus dollars in federal funds to wildlife agency coffers. The money would be earmarked for conservation-oriented projects. "Which would alter that dynamic."

We skim our menus. At the next table, they're on Miley Cyrus. ("She's amazing." "She *is* amazing.") Breck is constitutionally immune to the distraction. On the drive over, I had asked about Aspen celebrities. I got: "Jack . . . Nicholson. Nicklaus? Which one is the golfer?" He knew that Kevin Costner has a place, because Kevin Costner once had a bear problem.

Breck puts his menu down. "The thing that's not being talked about enough is this. You have a recovering bear population that was suppressed mightily at the beginning of the last century." The prevailing attitude toward America's wildlife in the early part of the twentieth century had changed little since settlers first crossed the divide. The first to push west were ranchers, subsistence farmers, cattlemen, fur trappers. Wild animals were either a commodity or they were varmints. Bounties were widespread. Bears were routinely poisoned, up through the 1970s. "We wiped out everything," Breck says.

The government was there to help. Breck's employer, the National Wildlife Research Center, has had many incarnations and names over the past hundred and fifty years but always one goal: effective, cost-efficient wildlife damage control. Whether the wild animals were predators taking livestock or birds and rodents helping themselves to crops; whether the name on the door was Division of Economic Ornithology and Mammalogy or Eradication Methods Laboratory or Division of Predatory Animal and Rodent Control, the goal was to help the rancher and farmer. What looked like pure wildlife biology — studies of animal behavior, food habits, migratory patterns — was biology in service of prosperity.

With the birth of environmental and animal welfare movements in the 1960s and '70s, a national conscience began to emerge. Activists pushed back against practices like den-shooting and airdrops of strychnine-laced bait. In 1971, Defenders of Wildlife, the Sierra Club, and the Humane Society of the United States sued to end the use of poisons in predator control. The following year, the Environmental Protection Agency (EPA) canceled registrations for strychnine and two other predacides. Advocacy groups sparked a shift in public sentiment, which over time has become impossible — and impolitic — to ignore.

A growing percentage of Americans feel a strong emotional connection to wild animals and disapprove of their destruction for economic reasons. In 1978, three thousand Americans were asked to rate their like or dislike of twenty-six different animal and insect species. In 2016, researchers at Ohio State University readministered that same survey. Compared to the first go-round, the proportion of respondents who reported liking wolves and coyotes was up by 42 and 47 percent, respectively. (The cockroach had also risen in popularity, from most despised — that honor now going to the mosquito — to second most-despised.)

Which is a gassy way of saying: bears are back. To the point where they're starting to get all up in people's worlds. "This is new territory for wildlife biologists," Breck says, forking salad greens. "And we're not very good at it. When I was an undergrad it was all about, How do we bring these populations back? How do we count them, manage them? Now it's all about human-wildlife interactions. How do we manage *this*? We're seeing wildlife biologists going . . ." Breck mimes banging his head against the table. "The game has changed."

At the moment, it feels unwinnable, this game. There are more bears, more wolves and coyotes, and ever more humans moving into their ranges. And no cultural consensus on what should be done when one of those animals ransacks a kitchen or kills some sheep or nips a steakhouse owner's backside. We have human-wildlife conflict and human-human conflict. We have ranchers and farmers and animal lovers hating on each other in a cultural clash that can feel as dug-in as the politics in this country. *Kill them all! Don't harm a single one!*

Breck and other specialists in human-wildlife conflict are starting to move their focus from animal biology and behavior over to human behavior. As a science, it's called human dimensions. The goal, stated unscientifically, is to find pathways to compromise and resolution. It often begins by bringing together people who aren't normally in the same room and getting them to listen to and even empathize with one another. Breck recently co-founded the Center for Human-Carnivore Coexistence. In early 2020, the center organized a two-day gathering of hunters, trappers, ranchers, and representatives of conservation and animal welfare groups to talk about reintroducing wolves in Colorado.

Breck came away hopeful. By the end of the second day, he was hearing people talk without hostility and in ways that struck him as productive. "The question now is, What happens when everyone goes back to their own corners?" Breck's hope is that whatever path the state pursues, it will be a decision reached by a group like this, rather than a few legislators in a back room.

Lately Breck has been spending time with Zach Strong, director of Carnivore Conservation at the Natural Resources Defense Council. The NRDC's more standard approach to Wildlife Services has been to sue them. Breck encouraged Strong to forge a relationship with the Wildlife Services director in Montana. Partly as a result of this unlikely pairing, three nonlethal — or "wildlife conflict-prevention" — specialist positions were created within Wildlife Services, two in Montana and one in Oregon. By showcasing the effectiveness of these hires, NRDC and Defenders of Wildlife were able to secure federal funds for hiring and evaluating similar positions in ten other states. Breck is hopeful that the new developments at Wildlife Services signal a shift within the culture there. Meanwhile, the Fish and Game Department in Idaho still funds a hunter- and rancher-friendly nonprofit that offers bounties on wolves.

All the government agencies agree on one thing: the fate of a wild animal that kills a human. Might even this someday change? Are there places in this world where the consensus favors sparing the animal's life, particularly in the case of a defensive attack? Bear biologist Dave Garshelis has been spending time on the Tibetan plateau, where brown bears routinely break into the homes of herders who have left for the summer to graze their stock. "They'd come back and the house would be a wreck, completely demolished. But these people are strict Buddhists and they don't want any retribution." Garshelis told me about a conversation he had with the local officer who responds in cases of animal attacks. "I asked him, 'What if you were called to a situation where you saw a bear on top of a person, mauling a person? Would you shoot the bear?' And he said, 'I don't have the right to decide whose life is more important, the person or the bear.' "

In India, around five hundred people are killed every year in encounters with wild elephants. Government policy is to compensate the families but not, with few exceptions, to destroy the elephant. The state with the highest numbers — 403 deaths in the past five years — is West Bengal. Maybe some answers are there, too.

3

THE ELEPHANT IN THE ROOM: MANSLAUGHTER BY THE POUND

In India there's a thing called an "awareness camp." I first heard the term from a researcher at the Wildlife Institute of India who runs elephant and leopard awareness camps. I imagined a camp in the American sense, with bunk beds and marshmallows, and I tried to meld this in my head with large, dangerous animals. Naturally I wanted to go. As it turns out, an awareness camp is closer to a national awareness day. I've seen listings for dengue fever awareness camps, diabetes awareness camps, traffic safety awareness camps, and at least one Snoring and Sleep Apnea Awareness Camp, which sounds like our bedroom. They're informational gatherings with the aim of making people aware of dangers they may not know much about or prefer not to think about and offering guidance in how best to avoid or survive them.

Every December that researcher, Dipanjan Naha, locks up his office in Dehradun, slaps a Government of India ON DUTY placard on a hired four-wheel drive, and heads out on a sort of awareness camp road trip. This year his cousin Aritra is coming along as his assistant, and I'm in the car, too. We're starting in North Bengal — confusingly, a region of West Bengal — where each year, wild elephants kill, on average, 47 people and injure another 164. Forty-seven people per year, in an area the size of Connecticut. India's forest department has wildlife rangers that get involved in these cases, but they don't kill the elephant. A few rangers will be on hand at Naha's first awareness camp, in the village of Bamanpokhri. I'm eager to learn what it is they do.

Outside the car window, agriculture rolls by: tea plantations, marigold farms, neat rows of rice plants set in the earth like toothbrush bristles. Small villages punctuate the paddies and plots — homes of corrugated metal and thatch, a temple, a few open-front bodegas. Cows loiter in the roads and small black goats are parallel-parked along the sides, but I see no other animals, no logical setting for elephants.

Elephants! Naha assures me they're not far off. It's winter, the time of year when herds are on the move. They forage at night and sleep by day in patches of teak and lal— remnants of the forests that once stretched from the Indian state of Assam through North Bengal all the way to the eastern border of Nepal. This "elephant corridor" has since been fractured and diminished, first by the many sprawling tea estates planted by the British and more recently by military bases and settlements of refugees and immigrants from Nepal and Bangladesh. Ever more humans are coming into these forests to cut wood and graze their livestock, turning elephant habitat into human. In their attempts to cross the land, the elephants encounter barriers, dangers, dead ends. The corridor is now a pinball game. Herds can become isolated in a single pocket of forest. They become "pocketed elephants." Keeping elephants in a pocket is as ill-advised as it sounds. The gene pool stagnates and the population density spikes. Soon there's not enough food to support them. They wander into villages to eat what they can find, and what they find is people's crops and granaries. Hello, human-elephant conflict.

Aritra points out his window as we pass a turnoff. "Two kilometers up that road a man was killed by an elephant. A few days ago. Three people were working on the roadway. They ran when they saw the elephant, and one man was separated, and the elephant followed him."

This is hard for me to imagine. I grew up with Babar and *National Geographic*. Elephants were gentle and slow-moving. They wore spats and bright green suits. They were never something to fear. This has created a minor disconnect between myself and my hosts. Our first night on the road, we stayed at a government-owned bungalow in a teak forest just down the road from a sign marking an elephant crossing zone. The bungalow's cook said he saw one near the gate the night before we arrived. My reaction to this piece of information was to announce my intention to go for a walk. It was around 7:00 p.m., two hours past the time when elephants head out in search of food and Dipanjan Naha and his cousin stop going near the forest.

"Don't go very far, Mary," said Aritra. We were sitting on the porch, drinking tea under the geckos and moths. Aritra has a round head and a friendly, slightly giggly disposition. He slips easily from his role as Naha's assistant to the more familiar role of younger cousin.

Naha didn't like my plan, either. "Please

be very careful."

They looked at each other, and then they set down their cups and got up to accompany me.

We got as far as the railroad tracks at the end of the driveway. Naha said a few words about the history of the narrow-gauge railway in India. We stood for a few minutes, as though waiting for a train. Aritra toed the gravel between the ties. "Let's go back inside."

To better appreciate the danger of an unplanned elephant encounter, sit down with someone who investigates the deaths. Saroj Raj is the range officer for the Bamanpokhri Beat of the local forest division, where every year since 2016, someone has been killed by an elephant.

Officer Raj has come to Bamanpokhri's blue-walled, one-room community hall — the site of today's awareness camp — to talk to people and answer questions. So far it's just me asking. The people who show up on time for an awareness camp, it seems, are the people who have to. Today that's a group of schoolchildren in plaid uniforms and a half dozen rangers from local wildlife squads. Naha isn't bothered. It's Diwali — a holiday week — and it's after lunch. "So they are sluggish to come."

Officer Raj gives me the particulars of the most recent fatalities. He begins, each time, with the exact date. You get the feeling there's a lot of paperwork. "Thirty-one October, 2018. Three workers in the road." The spot we passed earlier. "Suddenly one elephant appeared."

One can be scarier than many. Herds comprise females and young elephants, the peace-loving jumbos of my childhood. A loner is typically male, and males can be trouble. Bull elephants go through a periodic hormonal tumult called musth, during which their testosterone levels are as much as ten times higher than at other times. This gives them a competitive edge with other males and with the herd's dominant matriarch, but also a degree of random volatility. The musth mood state ranges, in the words of Asian elephant specialist Jayantha Jayewardene, from "hyperirritability" to "a charging or destructive tendency" toward other elephants, human beings, "and even inanimate objects." Villagers know this. "The men tried to escape into the bushes," says Officer Raj. "One person unfortunately fell down."

Officer Raj supplies the detail Aritra was careful to omit. The elephant stepped on the man's head. When you weigh 6,000 pounds, simply stepping or kneeling on a person — or, in the 1992 case of a riled circus elephant named Janet, performing a headstand upon the person — is an effective means of killing. But Officer Raj's determination in this case, based on footprints and disturbances of the surrounding vegetation, was that the death was accidental.

"Sixteen October, 2016. Also an accident." A man was climbing up a riverbank when he encountered an elephant. "It was slippery," recalls Officer Raj. "Both slid down the bank. The elephant just rolled over him." Elephants sometimes kill the way cars kill: by being large and running into — or over — something much smaller. (Elephant keepers try not to get between a wall and an elephant.)*

* At the zoo where I worked in my twenties, the pay rate for the elephant keeper was slightly higher than that of workers who cared for other animals — though not because of the risks. She was paid a "shit differential" because she had to shovel excessive amounts of it. Well earned: according to a 1973 *Smithsonian Contributions to Zoology* document, an Asian elephant defecates 18 to 20 times a day, dropping "4 to 7 boluses" of around 4

"These elephants were not in the intention to kill," Officer Raj says. How does he know? Because the bodies were in one piece. "If an elephant is in an angry mode, the body will not be intact. It will get in pieces." A book by Jayantha Jayewardene includes a list of the nine recorded methods by which an angry or musthaddled elephant has killed a human being. "Placing a fore foot on one limb of the victim and ripping off the other with the trunk" is number 3. (Elephants use a similar anchor-and-pull technique to strip an uprooted shrub of limbs and leaves to eat.) Rulers of Ceylon (now Sri Lanka) in the 1600s were said to have taken advantage of this natural behavior, training elephants to serve as executioners. An engraving from *An Historical Relation of the Island of Ceylon* depicts an elephant so engaged. One forefoot rests on the malefactor's torso, and the trunk is wrapped around the man's raised left leg. Were it not for the caption ("An Execution by an Eliphant") and the raggedly torn-off arm in the foreground, you could get the impression that Ceylonese monarchs trained elephants to serve as bodyworkers.

---

pounds each, for a daily output of 400-plus pounds.

Naha's awareness presentation will stress the importance of not upsetting an elephant. This is summed up by item 20 on the Best Practices for Mitigating Human-Elephant Conflict poster Aritra has just hung up on the wall of the community hall: "No Rambo-Style Action!" Shooting an elephant is not only illegal but, depending on the caliber of weapon, pointless. Our Janet withstood fifty-five shots from Palm Bay, Florida, police officers' 9mm duty revolvers as well as the first round fired by an off-duty SWAT officer with a stash of ammo designed to penetrate armored personnel carriers. (The second round stilled her.)

The safe thing for a villager to do upon sighting an elephant or a herd in the area is to call the 24/7 hotline and let Officer Raj's elephant squad handle it. The squad knows that elephants are social animals, so they stay calmer if they're herded off the premises in a group. The rangers converge on them from the sides, like cowhands, and move the whole group in the direction of the forest out of which they've come. By now, the animals recognize the sound of the elephant squad vehicles. "We drive into the area, and they go." Officer Raj smiles slightly. He's not a smiley guy. "This is a convenience for us."

Officer Raj makes elephant patrolling sound like mall security, but apparently it's quite risky. A beat officer sitting with us has been charged four times. "They tell you not to run," he says. "And I tell you, this is very hard to do when an elephant is coming straight at you!" My request for a ride-along is turned down, as is a second request. The expression on Officer Raj's face suggests I've become a "nuisance event," so I leave it.

Deaths, when they happen, tend to occur in the half hour or more it takes for the squad to arrive. Upon discovering elephants raiding their crops, villagers rush out of their homes, yelling, throwing stones, lighting torches and firecrackers.* A village may have freelance "elephant chasers" wielding

---

* That elephants are reliably spooked by fire and sudden loud noises limited their usefulness in war. Though the sight of armor-sided "elephantry" with swords lashed to their trunks conferred, from a distance, a psychological advantage, this quickly evaporated as the two sides drew closer. Records exist of elephants turning and breaking ranks at the sound of musket fire or the sight of flaming arrows. A fleeing, sword-waving elephant storming its own battalion likely racked up as many casualities as it would have inflicted on the enemy.

spikes and carrying out other non–Best Practices. Bulls and dominant matriarchs may charge in defense, and normally placid females and calves may panic and stampede. In the dark of unlit fields and paddies, people stumble and fall and elephants are running blind and, as my mother liked to say, *somebody's* going to get hurt.

"The elephant we can guide easily," says Officer Raj. "To guide the people is the hard part. They are not in the condition to listen." They're upset, and that is understandable. Village farmers work hard and have little to show for it. A single Asian elephant may consume three hundred pounds of vegetation in a day. Between the raiding and the trampling, a small herd can quickly torpedo a season's labor and livelihood.

An elephant among the crops is a powerful impetus to unwise action. Throw in the wobbly judgment and dimmed impulse control of inebriation, and the results can be dire, Naha says. He squats in front of a speaker, untangling a spaghetti heap of wires. "This is what we see. A group of people are drunk. Someone wants to be the hero, so he goes in front of the animal, harasses it, and that animal in self-defense . . ." Naha, too, avoids the verb *kill*, with its undertone of intent. "There is an accident." By his own data, 36 percent of the people killed by elephants in North Bengal between 2006 and 2016 were drunk. Later, I would see this headline in the *Hindustan Times:* "Drunken Man Challenges Elephants' Herd, Trampled to Death in Jharkhand." (Jharkhand borders West Bengal.) "He tried to fight with them," a forest ranger told the reporter. "Them" was eighteen elephants.

Dangerously, an elephant also enjoys a snort. In North Bengal, elephants drink what the villagers drink: *haaria,* a home brew fermented and stored in quantities sufficient to inebriate an elephant. (Because elephants lack the main enzyme that breaks down ethanol, it takes less than you'd think.) According to Officer Raj, two things happen when elephants liquor up. Most just stumble away from the herd and sleep it off. But every herd seems to have an aggressive drunk — the matriarch, often, or a bull in musth. Whatever you do in this life, stay away from an inebriated bull elephant in musth.

There is data to support Officer Raj's observations. In 1984, as part of a study at the UCLA Department of Psychiatry and Biobehavioral Sciences, three Asian elephants "with no known history of alcohol use" and seven African elephants from Lion Country Safari were served a "large calibrated drum" of grain alcohol-laced water. The animals tended to wander off, away from the herd. They stood or leaned with their eyes closed or "wrapped their trunks about themselves." They skipped meals. They didn't bother to bathe. The matriarch became louder and more aggressive, as did a bull named Congo.

To discourage elephants from tippling, villagers may drag the hooch indoors. A terrible idea. Because now it is not drunk elephants they need to worry about, but elephants determined to become drunk — that is to say, elephants who smell booze inside the house and see no reason not to knock down a wall to get to it. In Naha's survey, 8 percent of North Bengalis killed by elephants were sleeping inside their homes when it happened.

The chairs are mostly full now, and Naha takes up the microphone to begin his talk. He speaks in Hindi, of course, so Aritra translates, intermittently leaning over and cupping a hand to my ear to deliver a stand-alone sentence in a decisive, oracular cadence. "Being *drunk,* you should not go in front of the elephant." "Always drive by the *back* side of the elephant."

Naha is an expressive speaker with a lexicon of fluid hand gestures. I didn't necessarily expect this. Offstage, there's an economy of movement and words. He stands with shoulders squared and his feet angled outward as if to create a stable base. He's like the man I keep seeing in the Ambuja Cement billboards, standing firm and unperturbed with an entire thundering hydroelectric dam in his arms. I've heard him say "I was once chased by a tiger" in the same workaday tone in which you or I might say "I was once in Omaha."

The philosophy behind an awareness camp is simple. If you want to get through to people, talk to them while they're relaxed and clearheaded. Sit them down and let Aritra serve them a cup of tea and a samosa. The more people understand about the biology of elephants and the behavior of herds, the safer the encounters. It mostly boils down to staying calm and giving the animals space. Especially mothers with calves, and even more especially, lone males, and extra-strength especially, males in musth. (Some hallmarks of musth, courtesy of Jayantha Jayewardene: profuse oozing from glands on the temples, frequent erections, and "fully opened ogling eyes with

99

roving eyeballs.")

There is another point worth emphasizing with people. Officer Raj mentioned it earlier, when we were chatting: "We the human beings, *we* are disturbing *them.*"

Awkwardly, this includes the actions of professionals like Officer Raj. Herding elephants into the nearest forest patch provides immediate benefits for the local villagers, but in the long run, Naha told me later, it aggravates the problem. Because it aggravates the elephants. They begin to associate humans with the anxiety and privation of being chased off when they're trying to eat. They start to stand their ground. There are reports coming out of conflict regions in the neighboring state of Assam that say female elephants are starting to become as aggressive as bulls.

A better system, Naha believes, would incorporate sensors to detect the approach of a herd. A warning would go out to village heads and trained local response teams who would monitor the situation and try to intervene before crops are trampled and chaos erupts. Naha doesn't mean motion sensors or heat sensors, both of which would be triggered by other mammals. He means seismic sensors: sensors triggered by vibrations so powerful only the footfall of an elephant (or a small earthquake) would have created them. Meanwhile, you do what you can to reduce the heavy footfall of humanity: keep working to restore forests and set aside preserves.

The assistant manager of the Gopalpur Tea Estate wears long, trim shorts of suit fabric and bulbous neon athletic shoes. His head is set at a backward tilt that makes him seem aloof, or maybe just in need of a new glasses prescription. He came to greet us as we drove in. An awareness camp for the workers, our second stop this week, is scheduled to begin in half an hour. But first, a cup of tea.

The assistant manager is a numbers man. The estate covers 1,200 acres, he says, placing cups in front of us. Two thousand one hundred workers collect the tea leaves. We listen and sip, and then he leads us back out the door and across the way, to an open-walled pavilion where Naha will be speaking.

The workers have already arrived. They look through the handouts that were placed on each chair inside a reclosable clear plastic bag (brand name: My Clear Bag). The women sit on one side, segregated from the men. Naha fiddles with the sound system, a holdover from the days before movie streaming, when the estate brought in live bands for weekend entertainment.

It's hot and thickly humid. The tea estate managers are late. The workers fan themselves with Their Clear Bags. Time passes. Someone from the assistant manager's house brings a tray. More tea! Ours is served in china cups with saucers. The workers sip from paper — tiny cups no bigger than a shot glass. *WTF*, I want to say, *you've got twelve hundred acres of tea.*

Here come the managers! SUVs arrive like squad cars, pulling up fast and braking hard. Doors open and slam, and the managers, five total, stride onto the scene. The workers rise from their seats in unison. Rather than joining Aritra and me in the audience chairs, the managers ascend to a row of tables on the stage. They crack open the water bottles that have been set at their places along with notepads and pens. There is a spectacular variety of mustaches up there on the stage.

The managers take turns with the microphone. Aritra yells his translations over the blast of dance hall speakers. The first manager urges the workers to pay attention, because we are near a river and a forest, and so we have elephant problems. He hands off the mic to the manager beside him, who outlines the estate's present elephant deterrence strategy: a team that patrols on a tractor, lighting firecrackers as needed. The microphone continues its travels. The next manager is another numbers guy. In the twelve years he has worked here, seven or eight people have been killed by elephants.

At last the microphone is handed to Naha. The workers do in fact pay attention, so closely that I wonder whether there might be a penalty for looking away. The managers whisper with each other and peek at their phones in their laps. One takes a call, holding a hand over the device as if, like a burp, the behavior could thereby be made less rude.

When the presentation is over, Naha invites the workers to join in with questions and comments. One tea collector immediately rises from her chair. She is older than many, maybe fifty, and is dressed, like all the women, in the colorful, patterned sari she wears to work the tea plots. Aritra jumps up to bring her the mic, but she doesn't need it. Her anger is an amplifier. The mustache showroom shifts in its seats.

Aritra resumes translating. "You tell us to change crops," the woman says, referring to

103

the small gardens kept by the workers. "From corn or rice to something like ginger or chilis, which elephants don't like. But we grow corn and rice to feed ourselves. Also, once the tractor passes, the elephant is coming, coming, coming again. Elephants need a lot of food." The woman sits down again. "We need different measures."

She's right, but good solutions are elusive. Measures that seem intuitively obvious are, in practical fact, limited — by their expense and by the new problems they create. Electric fencing is a for-instance. You need enough of it to keep the herds out, but not so much that it blocks their migrations. Maintaining and repairing long stretches of fencing is time-consuming and costly and often doesn't get done. Or is done wrong. The voltage has to be high enough to discourage an elephant but not so high as to electrocute it. On average, fifty elephants a year are electrocuted in India.

And there is the considerable challenge of elephant intelligence. An Indian elephant, confronted with an electric fence, will likely soon figure out how to get past it without getting shocked. It will figure out that wood doesn't conduct electricity. It will push down the posts or it will pick up a log and use it to press down the wires to let others step through.

The elephant's intelligence has not always served it well. Elephants have been put to work — by the Indian military, historically, and more recently by the timber industry. They're treated like employees, in that the forest department keeps a log of their hours on a work register. These "duty elephants" are of course not paid a salary but, Naha told me, at age fifty they receive a "pension," in the form of retirement lodgings at a *pilkhana* with meals and a daily bath followed by a rubdown with oil.

As people get up from their chairs to leave, I ask Aritra to introduce me to the woman who dared to speak out. Her name is Padma. She has reason to be upset. A week ago, she was awakened at 4:30 in the morning and saw that an elephant had smashed through a wall and eaten the grain she sells in a small bodega in the worker colony. She has yet to receive the compensation owed to her from the tea estate.

A subset of managers hovers within earshot. One sidles up to derail the conversation. "Hello, you are from America, my son works in Memphis, in the best hotel, do you know the Memphis hotel ducks?" I actually do know this hotel, these ducks that descend the lobby staircase for no discernable duck or hotel reason. I keep my focus on Padma.

Aritra continues translating. "This is the second time it has happened to her."

Still playing on the manager channel: "At five o'clock the ducks come down . . ."

Naha joins us. He suggests we drive over to the workers' colony to see the wreckage of Padma's shop. The managers exchange urgent looks, but it's too late. We squeeze into our car with Padma and back away.

The tiny bodega isn't so much ransacked as flattened. A wall of corrugated metal lies crumpled beneath a concrete support beam. On another occasion, an elephant broke into Padma's home while she slept. This is a place where "the elephant in the room" is not a metaphor, where elephant jokes are no joke. What time is it when an elephant sits on your fence? Probably around 11:00 p.m.

Elephants are vegetarians, but they are not picky eaters. They'll eat most parts of a plant — grains, grass and leaves, stems, twigs, bark. On a tea estate in the Sonitpur district of Assam in 2017, three wild elephants broke into a workers' shop at 2:00 a.m. and helped themselves to the cotton fiber product known as rupees. They broke open the cash box and consumed 26,000 rupees in large denominations.

One thing Indian elephants won't eat is tea leaves. Everyone here likes to drink tea, but few, human or beast, like to eat it. The leaves are too bitter. Small crop losses result when elephants trample plants as they pass through an estate, but, by and large, it is the workers, not the owners and managers, who suffer.

Nonetheless, Padma and her neighbors say they feel no anger for the elephants. Seventy-five percent of the people Naha interviewed about elephants that come into their villages report positive feelings for them. Given the amount of death and destruction caused by elephants in North Bengal, retaliatory killings are surprisingly rare. Naha says just three to five elephants a year.

With Aritra's help, I tell Padma what happens to large mammals in the United States that injure humans or break into homes. I ask her whether people she knows ever express a desire to kill an elephant that has broken into their home or shop. "Why would you kill a god?" she says, referring to the elephant-headed Hindu god Ganesh. "We just say, 'Namaste and please go away.'"

Padma leads us out to the tea field currently being harvested. Leaf collectors are spread out along the rows, working the waist-high bushes. Only the new, tender bright green leaves are plucked. The workers remind me of steel pan drummers, standing still with their arms moving crazy quick. They are fast because they have to be. If they don't meet a set quota, their pay is docked.

Naha bends down to show me the empty space below the tea bushes. The leaf pluckers sometimes surprise a female leopard resting in the shade with cubs. The animal may awake in a panic and, if it feels cornered or threatened, jump at the worker. Deaths are rare, but injuries happen. Ninety percent of leopard attacks in North Bengal happen on tea estates.

We watch the women work. As soon as their hands are full, they reach behind their heads to stuff the leaves in a cloth sack slung over their foreheads. Now I understand why Padma, earlier, had shown intense interest in my backpack. You'd think the managers could spring for bags more ergonomically suited to their task. I say this to Naha.

He rolls up his sleeves. "They are paid one hundred fifty rupees a day." This is less than the cost of a cappuccino at the Delhi airport. "It's a colonial mentality. They are the same tribal laborers that the British brought in from central India. They use them because they are considered hard-working and obedient."

When Naha told me about the work register for animals, about pensioner elephants and their bathing regimens, I was initially impressed. Here is a government guaranteeing working animals some of the same benefits normally accorded human workers. Seeing, now, how the tea collectors are treated, are legally permitted to be treated, it's less uplifting than it had seemed. Depending on your species, religion, gender, and caste, India may be a better place to be an animal than it is to be a human. In 2019, the government of Delhi announced plans to revamp one of the five sanctuaries it maintains for the city's free-ranging, traffic-snarling, sacred cows. Perhaps in response to criticism that the city provided better care for its cows than for its citizens, the city's Minister of Animal Husbandry announced, "We are planning a unique coexistence programme where elderly will be allowed to stay with the cows."

Things have lately grown more extreme. The current prime minister, Narendra Modi, rose to power on a surge of Hindu nationalism. This is a man who bestowed personhood status on the Ganges. A river enjoys human rights protections, while women like Padma earn 150 rupees a day and Muslims are lynched for selling beef.

As we get in the car to go, the assistant manager trots over with an armful of Mylar bags: a pound of tea for each of us. Aritra thanks him, then turns to Naha as the car gets underway. "It's CTC." (The acronym describes a processing method.) To me he explains, "The cheapest kind."

Wildlife is the theme at the government-owned Jaldapara Tourist Lodge, where I'm staying tonight. The grounds feature life-size plaster models of local wildlife. A number of the figures have been knocked over or have broken-off appendages that lie where they've fallen, on the incongruously well-manicured lawn. The whole courtyard has the feel of a miniature golf course after a party of drunken golfers played through, beaning the statuary with errant balls or maybe their golf clubs.

I love this kind of place, love the surreal decay of it, love the clerk who does not know where breakfast is served or even *if* breakfast is served, love everything, really, except the rat turds on my balcony. I tried to imagine what could be the draw, for a rat, of a balcony with no food and nothing in which to nest and not even much of a view. It appeared to be simply the place rats go to have a dump. The Jaldapara rat toilet.

The lodge is scheduled for a remodel, because the West Bengal forest minister is working with the West Bengal Tourism Development Corporation to create a rhino preserve out of the neighboring forest. This happens to be the forest to which Naha, Aritra, and I are shortly headed. We're going to track a radio-collared "conflict leopard" — leopard 26279 — that was relocated here a year and a half back, when the land was earmarked to become a wildlife preserve. The animal was not full-grown then, and Naha wants to check to be sure it hasn't outgrown its radio collar.

While he's here, Naha will talk to villagers who live on the edges of the forest. Radio-collaring provides points on a map, but it doesn't answer the questions that ought to be asked when a predator is released into people's backyards. Has the leopard been taking your goats? Are people okay with it being here? Naha has been following up, like a social worker on a foster care placement. He's been tracking this leopard remotely and phoning alerts if he notices it moving toward their homes.

This morning we have a new driver,

Ashok. He rarely joins the conversation, preferring to give his full focus to the driving. This is a switch from the last driver, who, to keep himself from nodding off, suction-cupped his smartphone to the inside of the windshield and began streaming a sitcom. (Naha was unalarmed. "One eye is on the phone, and one eye is on the road.")

We turn off the paved roadway onto a dirt track through increasingly thick underbrush. Branches scrape the sides of the car. Ashok has grown quiet. He seems tense. Did I say something wrong? Is he worried about scratching his paint job?

Just past a cluster of homes, we pull up alongside a man dousing his cauliflower patch with pesticide from a backpack tank. Naha gets down from the car, and the man shuts off the nozzle. One eye is cloudy. Three other men wander over. Aritra listens in. Nothing to report. They haven't seen the leopard in a while.

We drive on. We pass a watchtower used by anti-poaching squads. Naha asks Ashok to stop so he can climb up and get a better signal. Aritra tells me to stay inside the car with him.

Naha descends the tower stairs and gets back in the car. He reports that we're now just a thousand feet from the leopard.

We drive on. The road dead-ends at a wide river. Naha gets out again. He walks along the sandy riverbank holding an antenna above him like a torch. On the other side, a group of men are waist-deep in the water, clearing an overgrowth of water hyacinth.

Naha leans in the car window to let us know we are now 500 feet from the leopard. *Dut-dut-dut-dut,* goes the receiver. We can't get closer, because the cat is on the other side of the river. He points. "Just beyond where those men are."

There's no bridge near, so we turn back. On the last few miles of the drive, Ashok breaks his silence. He and Naha speak, in Hindi. After Ashok drops us off, I ask Naha what they were talking about.

"His father was killed by a leopard."

The father had gone to collect firewood. When he didn't return home, Ashok, who was twelve then, went with some friends to look for him. They found him, still alive, in a riverbed. The kind of place where the men were working today, a few dozen feet from a leopard. "They brought him to the hospital," Naha says, "but he couldn't recover. He sustained a lot of injuries. His eyes, and all that. He must have been very badly mauled."

Ashok will not be our driver for the next leg of our trip, and this is for the best. We are headed for Pauri Garhwal, a hot spot for leopard attacks. These are not the kind of encounters that take place on tea estates when a worker surprises a leopard sleeping under the plants. These are stalkings, killings undertaken with intent.

4
A Spot of Trouble: What Makes a Leopard a Man-Eater?

The road to Pauri Garhwal carries travelers through the Middle Himalaya — the sparsely settled, mildly mountainous realm between the low foothills and the airless white monsters of the Greater Himalaya. It's a lovely drive but, as road signs will say to you, HIGHLY ACCIDENTAL PRONE. Landslides happen so often that the sides of some mountains, at a distance, look like ski areas. The higher you drive, the steeper the slopes and sharper the curves, to the point where you're forced to take each bend blind, honking and bracing for impact.

The road traces the old pilgrimage trail that links the Hindu shrines along the Ganges river. In past centuries, devotees walked the trail barefoot and slept in simple thatch shelters. The danger then was not crashes but leopards, which had a documented proclivity for slipping through an unsecured doorway. Between 1918 and

1926, government recorders ascribed 125 kills to a single leopard, known in the global media of the day as the Man-Eating Leopard of Rudraprayag.

The shrines remain and the people still come, but now they drive and book hotels. Modest accommodations line the road: Hotel Nirvana, Om Hotel, the unrelaxing-sounding Shiv Hotel. Our driver today, Sohan, is friendly and seemingly imperturbable, steady in the face of all that India drops in our path: the cows, the rockslide rubble, speeding motorcyclists, a dilapidated loom. I've seen the serenity lapse only once, at the sight of a man urinating over a guardrail. Sohan drives so competently that I've stopped fretting about the nonfunctional seat belts in the backs of Indian cars, a preoccupation Naha finds amusing. (Naha on airbags: "You mean that balloon that comes up?")

We'll be visiting three mountain hamlets today and tomorrow, all of them leopard attack hot spots. If there's time, we'll go to Rudraprayag, which has grown by now to a small city. A monument stands on the spot where the famous man-eater was killed by the famous man-eater hunter Jim Corbett. Naha made a pilgrimage of his own last year, to look for descendants of Rudraprayag villagers who may have been alive then. He leans into the back seat to show me photographs. The Corbett monument needs repair. The pedestal is cracked, and the great hunter's mustache is chipped. Naha managed to track down a grandson of the village priest Corbett worked with. One of the things the man told him was that when a leopard stalks and kills more than three or four people, villagers consider it a demon.

I don't buy in to demons, but I do wonder what's going on with these leopards. In North Bengal, where we've just come from, leopard attacks are inadvertent encounters. After a quick scuffle, the surprised cat takes off. There may be injuries, but rarely are there fatalities. Here in Uttarakhand's Pauri Garhwal district, leopards stalk humans as prey. Every year, in an area smaller than Delaware, leopards typically kill three or four humans. Between 2000 and 2016, Naha reports, leopards attacked humans 159 times. A large majority of these attacks were predatory, he says.

What causes a species to update the menu? What happened in Pauri Garhwal?

Corbett blamed the flu pandemic of 1918. So many people died so quickly, he wrote, in *The Man-Eating Leopard of Rudraprayag,* that the Hindu funerary custom of carrying bodies to the Ganges for cremation was for a time replaced with a more expedient rite. A burning coal was placed in the mouth and the corpse cast down the hillside in the general direction of the river. A leopard will readily scavenge a meal, and Corbett surmised it was these bodies that gave the Garhwal carnivores their taste for human meat. Likewise, the man-eater of Panar — another Jim Corbett project — began its killing spree in the wake of a cholera outbreak. Corbett claimed this animal had killed four hundred humans by the time he stepped in. (The figure is questioned in *Leopards in the Changing Landscapes,* by H. S. Singh, formerly of the Gujarat state forest department, and by others, including Naha. Corbett stalked book sales as skillfully as he stalked big cats.)

One thing that may work in favor of the North Bengalis is that their leopards were heavily hunted by the British Raj and their royal Indian cronies. Singh wrote that between 1875 and 1925, hunting parties killed 150,000 leopards. "They may," Naha says, "still regard humans with fear." Now that my state's mountain lions (aka cougars)* are no longer bounty-hunted, are

* "Mountain lion," "cougar," and "puma" are

they becoming less wary of humans? I posed the question by email to California mountain lion researcher Justin Dellinger (whom you'll shortly meet). He didn't think so. Mountain lions are not so much wary as stealthy — a trait that evolved over eons, probably because it made them successful hunters.

The hills of Pauri Garhwal have a terraced, wedding-cake appearance. Terracing creates level spaces for hillside farming, but

---

regional names for the same species. Florida calls them "panthers," and in South Carolina they're "catamounts." The name "Rowdy" applied to just one, a cub captured by Clark Gable on a 1937 hunting expedition. Rowdy was one of two cubs Gable intended to bring back to surprise paramour Carole Lombard, who had jokingly told him to bring her back "a wildcat or two." According to Stanley P. Young, co-author of *The Puma,* Rowdy escaped the first night, in his new collar with the engraved nameplate (only to be bagged a year later, collar intact, by a mystified hunter). Rowdy's sibling was presented to Lombard and soon thereafter donated to the MGM Studios zoo. Lombard had earlier given Gable an enormous ham with Gable's face pasted to the wrapper, so the cougar appears to have been the victim of misguided gift-giving one-upsmanship.

I see no farmers and, when I look closer, no crops. Naha explains that the region has undergone a sizable out-migration. Villagers have left to look for work in cities, because almost anything is easier and more profitable than farming the hills. Terraces are difficult to irrigate, and when the crops mature, monkeys and wild boars show up to raid them. In the decade between 2001 and 2011, 122 villages were abandoned. You see it in the scenery: mile after mile of fallow terracing. It's like driving through a topographical map. In places, the contour lines have begun to soften as the native vegetation returns. This "rewilding" creates scrubland that serves as cover for leopards on the hunt. Nearly 99 percent of the villagers Naha interviewed for a paper published in PLOS ONE reported a belief that this has brought leopards closer to people's homes. Seventy-six percent of leopard attacks on residents of Pauri Garhwal happened in a spot with medium or dense shrub cover.

As the people have moved out, more livestock have been left to graze unguarded: easy pickings for leopards. Naha makes the point that chasing prey on steep slopes is, like farming on them, a challenge. Goats and calves make dinner a snap. Compared with deer and other natural prey, domesticated animals are slower and less wary.

As are human children. Forty-one percent of the Pauri Garhwal villagers killed in leopard attacks, Naha's survey shows, were between one and ten years old. An additional 24 percent of the fatalities were youth between eleven and twenty.

Here Sohan joins the conversation. Aritra is dozing, so Naha translates for me. "He has seen this." It was 1997. A thirteen-year-old girl was alone in a field, cutting grass with a sickle. It was around 4:00 p.m. Sohan was resting in a car a few yards away when a leopard appeared. "He saw everything in front of his eyes," Naha says. "The leopard attacked from behind. It jumped on her back and pierced her jugular. There was blood coming out. It was very harsh."

I ask Sohan about the physical condition of the leopard. Did it limp? Was it old or underweight? Jim Corbett, in another of his memoirs, put forward a theory about man-eating Bengal tigers that held that they were sick or injured, and that they went after humans because that's all they could bring down. (Like the cougar that attacked Ben Beetlestone.) "With the leopards of Pauri Garhwal," Naha says, "this has not been the case." He scratches the side of his face. His once-precise beard line is rewilding.

Sohan agrees. After the killing, the girl's family was persuaded to leave her body, briefly, where it lay. The leopard would return to its kill, and local hunters had been called. The body was chained to an iron stake, and the hunters waited. It was Sohan who drove the leopard's body to the forest department for necropsy. Other than the gunshot wounds, the animal had no injuries, and no teeth were broken or missing. "It was in perfect condition."

More often than not, Naha says, man-eaters are females with cubs to feed. In his book, Singh objects to the term "man-eater," because it suggests an animal "gone 'mad.'" It blames the leopard rather than the changes humankind has brought — the rapid disappearance of forests and the prey that lived in them. Besides, he points out, to a carnivore, meat is meat. "Big cats eat all varieties of flesh. . . . So why not human beings? Who gave these magnificent big cats this derogatory tag to begin with?" he asks. And then answers: Jim Corbett.

Naha's wife, Shweta Singh, also a wildlife biologist, has joined us for this leg of the trip. The couple met at the Wildlife Institute of India, where both are employed, but Shweta did not come along for her work.

She came because it's beautiful here and the air is cleaner, and because this week is Diwali and she'd like to spend it with her husband. Shweta is a little younger than Naha and lighter of spirit. She and he share a passion for research and the wild places it brings them to. She and I, for the colorfully packaged snack foods of India, bags of which you find at every roadside bodega, hanging in strips like linked sausages.

Sohan has pulled over in a small town. We're stopping for tea (and two bags of Masala Munch). Below the cafe window, a steep incline bottoms out at the Ganges. The river's glacial origins are evident in the water's chalky aqua hue. It's noticeably colder up here. The women wear cardigans under their saris.

Naha is filling me in on forest department policy for leopards that harm or kill humans. Unlike in the United States and Canada, where the animal would be, as they say, "lethally removed," here a distinction is made between defensive and predatory behavior — or as Naha puts it, provoked versus unprovoked. If a particular leopard is known to have killed and fed on three or more humans, only then will the chief wildlife warden of the state formally declare it a "man-eater," whereupon hunters or forest department staff may shoot it. How do they know it's the work of one leopard? They set up wildlife cameras in the area and learn to recognize the local cats and their turf. (They keep them straight by their spots. The pattern of a leopard's rosettes is unique, like a set of human fingerprints.)

Declared man-eaters are no longer translocated. Here the thinking mirrors that of North American wildlife agencies. If you move a leopard and it kills a person in the new location, now you, the forest department, may be held liable. The human-leopard conflict researcher Vidya Athreya reports that translocation itself makes attacks more likely. After forty leopards were moved to a forested area in Maharashtra in 2001, the average number of attacks per year jumped from four to seventeen — and not just because there were more leopards in the area. Athreya attributed the rise to two factors. First, the leopards had lost their fear of humans during the period of captivity leading up to the translocation, and second, the stresses of being captured and released into unfamiliar territory had made them more aggressive.

On top of everything else, translocation is, as with black bears, likely to be a temporary solution. Remove one leopard from its turf, and another soon takes it over. Often the new arrival is a subadult recently dispersed from its mother. That can mean trouble, in that inexperienced hunters are more inclined to go after easy prey.

If not translocation, what, then? What does a forest department do with a leopard that has killed just once, or is preying on livestock? Naha pulls his attention from a nature program on a TV mounted on the cafe wall above our heads. A narrator is sharing facts about shrews.

"It can be trapped and kept in captivity," Naha says. When I ask where, he uses the word *zoos.* I ask about the possibility of visiting one. "They are not open to the public."

So, not really a zoo. I'm trying to picture it. "Open spaces, like a fenced preserve? Or cages?"

Naha runs his hands through his hair. The side of his head has been styled by the four-hour blowout of an open car window. "Both. They are enclosed for some time, and then released into a bigger area for exercise."

"Like a prison." Naha doesn't argue with the comparison. The leopards do time.

Later, tooling around the internet, I found a master plan for an upgrade to what appeared to be a similar place — the South

Khairbari Rescue Centre in West Bengal, not far from where we were last week. Twenty-five "night shelters" opening onto a paddock area. It was not for conflict leopards per se but for rescued circus tigers and orphaned tiger and leopard cubs. This facility, too, was closed to the public.

The heading "Disposal of Solid and Liquid Waste" caught my eye. "Faecal matters" were treated differently than carcasses, which went to a Bone Collection Unit. "The bones so collected are disposed of by way of selling." Was the government of West Bengal taking part in the illicit trade in "medicinal" wildlife parts? Given the high price of tiger and leopard bone, I wondered about the temptation to change a life sentence to a death sentence.

As in the United States, there are people who object to the government's policies and take matters into their own hands. But where we have, say, Californians freeing bears from Fish and Wildlife culvert traps, here in Pauri Garhwal the anger runs in the other direction. Villagers want "man-eaters" killed, and they don't want to wait for a second or third victim. A mob mentality can quickly take hold. In one of the hamlets on our route, a leopard had recently killed two people. Residents did not contact the forest department but set a cage trap themselves. A young man who spoke some English brought us to the spot where the animal had been trapped. "There was too much angerness," he said. "So due to this angerness, he was burned, the leopard. In the cage itself." After a quiet moment that I mistook for melancholy, he held up his smartphone. "Can I selfie?"*

Shweta knows this case. Her forensics lab at the Wildlife Institute of India received the evidence and the remains — in this case, "ashes, and stones with blood." Forest department staff will endeavor to be sure they've caught the correct leopard, but villagers will kill whatever steps into the trap. "They don't know if they have the right animal or not," Shweta says. "They just want to take their justice." Had things been done properly, officials would have first tried to match the DNA of the trapped leopard with DNA found on the victim's skin or under the nails. (Linkage!)

Naha glances around the cafe. "This is

* Proximity to the border with Tibet/China means a heavy military presence, and that means military-sponsored cell towers and one of the many curious contrasts of modern India: villagers who have smartphones before they have indoor kitchens.

not a good place to talk about this." Though few here speak English, they know the L-word. This morning I noticed that the government ID placard is gone from the windshield.

A similar conundrum faces wildlife managers in American states that have three-strike or two-strike laws. If the state were to adopt a never-kill policy, ranchers would likely go back to killing the predators themselves — a practice known among wildlife professionals as "shoot, shovel, and shut up."

Again: too much angerness. "If you're a rancher, your life is sheep," Stewart Breck said, back in Aspen. "There's a strong emotional component." Breck has six llamas. You would not say that his life is llamas, but he will not forget the day he walked behind his house and saw two of the neighbor's dogs at the neck of one of his llamas. Here in the Himalaya, it's not just people's livestock being killed but also their family members.

Naha takes his jacket from the back of his chair. "Let's go."

By noon, we're deep in leopard country. Naha is pointing out the car window at the place where an eleven-year-old was attacked and killed walking home from school. The last five miles have been a stop-and-start narration of death, in Naha's streamlined monotone.

At a bus shed on a lonely stretch of road near the village of Kolkhandi: "One elderly man was sitting there, and he gets attacked."

On the road to Ekeshwar, our next destination: "Here were two attacks. An old lady, at five in the morning. And in the same spot, three years back, a person, thirty-eight years old. Returning from the fields."

Across from a field that abuts a forest in Maletha, near Ekeshwar: "Fifteen or sixteen people were there, all cutting grass. It was a most audacious attack. One lady was taken. Right in the daylight."

Sohan pulls onto the shoulder. The road to Ekeshwar is too narrow for our vehicle, so we'll park here. Naha pulls his bag from the back and shuts the door. He points to the hillside that slopes down to the village. "Here a lady was attacked and eaten. Late evening. 2015."

Along the half-mile walk, we pass a woman walking with a sickle in her hand. "See that lady," Naha says, tipping his head in her direction but not pointing, as though he's about to share some juicy item of village gossip. "She's going to the forest. She is taking a risk, going alone to cut grass." She has no choice. Winter snows are coming, and the cows will need hay to eat.

Local government in Pauri Garhwal takes the form of a "head system." Win the trust and support of the village head, and/or the priest, and your job is vastly easier. Naha has been back and forth to this area many times, creating relationships with both, and it's been paying off. We stop first at the home of Ekeshwar's village head. He is out, but we are welcomed by his brother Narender, a tall, gap-toothed man wearing flip-flops, one maroon and one gray, despite the chill. He invites us in, or rather up. It rains so infrequently this time of year that rooftops become living spaces. Red chiles have been spread in the sun to dry. A satellite dish is held upright by an arrangement of rocks.

Shweta translates for me. "He likes leopards even though they sometimes take his livestock. He says this is its natural prey, and he is acceptable to that. He does not support anyone who does killing." In the words of an American rancher I met last year who is also, improbably, a mountain lion activist, "When you have livestock, there's going to be some deadstock."

Naha has asked Narender and his brother

130

to nominate villagers for a wildlife response team. This would be the same sort of early responder team Naha has been helping to establish in North Bengal, but here it's humans, not animals, that the team tries to control. Most of the nominees are army veterans, because they are respected in the community and because they are, Naha explained, "capable of controlling a mob." I've seen the equipment list for the team members. It includes "polycarbonate riot shield of 3–5 mm thickness" and "fibre stick used by Police Department."

Shweta points out that people's ire is for the government as much as for the leopards. If there were school buses, children wouldn't have to walk two miles at dusk, when the risk of a leopard attack is greatest. If there were hospitals and ambulances, an attack might not mean a life lost. But there are not. A leopard is an expedient outlet for their anger.

Naha has held awareness camps at many of these villages. He encourages parents to have their children walk home from school in groups. He tries to discourage people from dragging their dead livestock onto the road for the vultures, because the carcasses also attract leopards. Attitudes and behavior change slowly in a small village like this.

Twenty years ago, Naha recalls, there were cases of Pauri women being nabbed by leopards as they squatted in the brush to relieve themselves at night. Indoor toilets were eventually built, but people wouldn't use them at first. "Slowly they are understanding it's okay to shit indoors."

Naha goes to check on a light he installed on an earlier visit. It's part of a controlled study to assess whether "fox lights" help keep leopards away from people's houses. These are solar-powered lights that turn on and off randomly to mimic the sight, at a distance, of humans patrolling with flashlights. It is shaping up to be an effective, if temporary, solution. To forestall habituation, the lights should be used intermittently. To get people to understand this has been difficult, Naha says. They want to leave them on continuously. Stewart Breck has had the same challenge with some of the ranchers who try fladry — fluttering ribbons tied to livestock fencing wire to spook coyotes and wolves. When they see that it works, they leave it up, rather than restricting its use to calving season and other periods of heavy predation.

This year Naha has been encouraging village heads to apply for funds from the Mahatma Gandhi National Rural Employment Guarantee Scheme. The money would allow the village to hire someone to keep the brush trimmed back from around the houses, and to build secure nighttime enclosures for livestock. These are the same suggestions that more progressive USDA Wildlife Services operators give to property owners who call because they want a mountain lion killed for preying on their livestock or pets. What if Wildlife Services made these things a requirement rather than a suggestion? Better yet, what if they arranged and paid for the brush-clearing, or for the enclosures to be built? What if non-progressive operators had to start being progressive? I ran this by Stewart Breck, by phone, when I got home. He did not see it becoming policy. "It's more of a philosophy."

I exhaled in a snorty, dismissive way. "That Wildlife Services pays lip service to?"

"Let's just say the ship is slow to turn. But it is turning."

The day ends high on a hill in the village of Khirsu, where the wildlife institute leases a house. It's unfurnished and unheated, but the view across the valley is a balm after the hours in a car. The slope directly behind the house is forested, and people have tied bunches of grass around the tree trunks to dry, out of reach of free-ranging cows. Aritra stands beside me on a balcony, intent on the hula-skirted trees. He hears something.

Naha watches his cousin. "Aritra is scared it's a leopard about to jump down from the hill." He points to a patch of sandy ground on one side of the house. "If there were a leopard around, there would be pugmarks there."

A large langur, black-faced and beefy as a baboon, drops from a tree and bolts across the slope. Scares the stuffing out of Aritra and me. "And there," deadpans Naha, "is Aritra's 'leopard.' " He goes inside to help Shweta fix dinner.

To enjoy the view (and because we have no table or chairs) we eat outside on a patch of concrete in front of the house. Shweta has built a cook fire and rigged it with a spit made from branches. Someone opens a quart bottle of Godfather Super Strong. By the time we finish eating, it is past ten. Shweta keeps the fire going. Naha is talking about the Aghori, Hindu monks who practice ritual cannibalism. His words are cut off by a bellowing outcry, louder and more jagged than a scream, human but barely. It's the sort of thing horror movie sound effects people aspire to but seldom attain.

"Leopard!" says Aritra. *"Leopard!"* He doesn't mean that a leopard is making this sound. He means a leopard is attacking someone. I know this is what he means, because it is absolutely the sound I, too, would imagine coming out of a person being killed by a leopard — terror plus pain plus the squeezing of vocal cords by clamping jaws. It's unfolding directly below us, at the bottom of the hill where the path from the village shops meets a cluster of houses. "Get up, get up!" We do, we get up, and we stand there, freaked and listening, trying to understand what's happening. Leopard? Insane person? Drunk? Aghori? Down below, other voices join, but they don't sound like people watching a leopard kill a neighbor or trying to stop it. Soon the voices trail off, as the afflicted entity is led or carried away.

It's late and the path is steep and unlit. We'll inquire in the morning.

We pack after breakfast and begin the precarious walk down the trail to the car with our stuff. Cook smoke rises from the houses below, and Himalaya morning sounds — women sweeping, men coughing, cowbells. At the bottom, Naha stops to talk to a woman standing in her doorway, to find out what was going on down here the night before.

He catches up with us at the car. It wasn't a leopard or a drunk. "It was a case of demonic possession," I hear him say in his tossed-off Naha way, as if it were no more remarkable than a sprained ankle.

Does this sort of thing happen often around here?

Aritra pushes a heap of sleeping bags deeper into the rear of Sohan's hatchback. "At least monthly," he answers after some thought.

I think back to what the priest from Rudraprayag said to Naha about leopards that kill three times — that they are demons. So maybe a leopard after all.

Our destination today is the city of Dehradun, home of Naha and Shweta and the Wildlife Institute of India. We are leaving behind the demons, but not necessarily the leopards. In 2009, an emaciated leopard showed up in Dehradun. Nineteen people were injured before the animal was shot.

Because most Indian leopards that visit a city do so at night, nabbing a stray dog or scavenging garbage and returning to the forest before daybreak, the excursions go largely unnoticed. It's when the big cat lingers past dawn that trouble sets in. H. S. Singh's book includes summaries of forty-three cases of "leopards straying in cities." They do the things people do. They visit temples and stroll college campuses and go to the hospital. One afternoon a leopard showed up at the Central Institute for Cotton Research. In a township outside Chandigarh, a woman came home to find a leopard asleep on her bed in front of a television program. In 2007, a leopard that had been seen for several days in and around the city of Guwahati was eventually trapped inside an upscale shopping complex, where it had been observed "prowling near an ATM," as though out of cash.

When things go well, the leopard is quickly tranquilized and released in nearby forest. More often, the story unfolds as it did in the case of the leopard who jumped the wall of the gated compound belonging to Bollywood star Hema Malini:

1. The person who first encounters the animal runs or hides. ("The gardener and the watchman locked themselves in a room.") In a common variation, it's the leopard that is locked in a bedroom or bathroom.

2. Police are summoned and, lacking tranquilizing darts and training, prove to be of minimal help. ("When one of the police teams tried to enter the house, the leopard growled at them, after which they waited for the forest department to arrive.") Though to be fair, some officers have proved to be skillful improvisers. Police responding to a call about a leopard in a suburban Delhi plywood factory hoisted a netted cricket batting-practice cage and threw it over the cat.

3. Long before the forest department arrives (a four-hour wait, in the Hema Malini case), the leopard makes a run for it. Malini's leopard escaped unharmed.

The Wildlife Institute of India sits on the periphery of a small forest, but Naha doesn' recall the Dehradun leopard. What they d have roaming around the institute ar rhesus macaques, dozens of them. And troop of macaque researchers, some o whom I'm planning to see tomorrow.

We're halfway back to Dehradun now Shweta has her earbuds in, nodding to he tunes. Aritra is trying to explain Hinduisn to me. As we drive down out of the hills, the black-faced blond-gray langurs we'd been seeing on rooftops and in trees are gone, and in their place are smaller, pink-faced monkeys with light red fur — rhesus macaques. These monkeys come right down to the road. They sit on the cement slabs that serve as guardrails, waiting for handouts or trash thrown from car windows.

Everyone in northern India has macaque stories. Naha awoke one morning to find one sitting on his chest. After a brief, intense stare-down, he yanked the blanket up between them and the monkey ran off. Another time, a macaque came up the back stairs of their apartment building and leapt onto the kitchen counter. "It could have just eaten some things and left, but no. He took the induction cooker and threw it on the floor. Just came in and did that and left. That made his day."

I've heard they'll snatch a person's sunglasses. "Yes," Naha says. "Or their cellphone. And then they drop it from the tree. Their main purpose in life is to harass people."

Shweta pulls out her earbuds. "They do this because this behavior is rewarded. When a monkey takes a phone, the person comes back and offers it food, because they know the monkey will take the food and give up the phone."

Naha isn't having it. "Shweta, remember they came to the terrace that time and flipped the flowerpot upside down?" He turns to me. "And then they'll go and shit right there. They feel good when you are harassed."

Shweta puts her earbuds back in.

Naha looks out the car window. "Definitely they do."

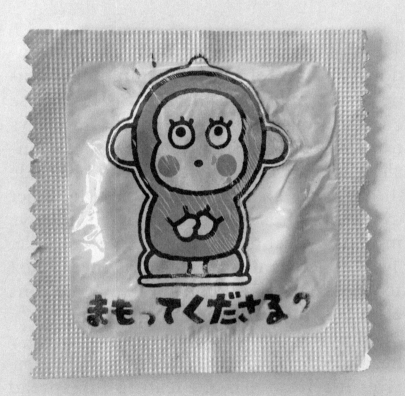

5
THE MONKEY FIX: BIRTH CONTROL FOR MARAUDING MACAQUES

The Wildlife Institute of India is a thrown-down cluster of concrete buildings connected by outdoor walkways. Because these corridors have no walls, rhesus monkeys from the neighboring forest can occasionally be seen walking along behind or beside the humans. Neither species pays the other much mind, as if the monkeys, too, have meetings to get to and photocopies to make. This nonchalant coexistence is in contrast to the state of human-monkey relations elsewhere in India.

"Simians Lay Siege to Agra," blares a headline in the *Times of India* the week I arrive. It's from a multi-feature full-spread special section, complete with signature two-color MONKEY MENACE graphic, the letter O taking the form of a fang-flashing monkey head. The lead story describes a baby fatally wounded after a rhesus macaque snatched him from his mother's breast. "Earlier this month," another *Times* piece states, "a bunch of monkeys had stoned a 72-year-old to death." The *National Herald* has had Agra's monkeys "in armies marching from one area to the other." During the eleven months I've had Google on alert for monkey news in Delhi and Agra, Indian newspapers have reported eight lethal macaque "attacks."

The past decade has seen a minor epidemic of people plummeting from balconies because of monkeys. I found news accounts of six deaths in the last three years alone. Most famously, there was the 2007 fall of Delhi deputy mayor S. S. Bajwa. While taking the air, Bajwa was startled by a group of macaques set on storming the house to look for food. As he tried to stop them — or get away from them (there were no eyewitnesses) — he lost his footing and hurtled over the railing.

While I question the hostile intent that the word *attack* suggests, a monkey home invasion is surely an unnerving experience. Each evening on a recent visit to Udaipur, I would sit in one of the city's many rooftop restaurants and watch langurs and macaques appear out of the dusk to begin their evening depredations. They sprinted up fire escapes and leaped from building to building like Tom Cruise in righteous pursuit. One night, making my way through an unmemorable dal, I looked up to see a langur vault onto a decorative beam above my table. Had the waiters not kept a stick close at hand for scaring off monkeys, the meal would quickly have become more memorable. When a forty-pound monkey abruptly drops in, you move without thinking. If you happen to be on a balcony or a roof, you too may shortly be dropping in from above.

The Wildlife Institute of India, according to several of these articles, is working on a contraceptive vaccine. The *Times of India* wrote of a shot that would " 'sterilise' the animal within minutes." Here is the dream! Easily administered, long-lasting birth control for overpopulous, problematic wildlife. Before leaving for India, I was unable to confirm an appointment date by email with Qamar Qureshi, the institute's director of research, and he has been out for the Diwali holiday. Naha agreed to escort me onto the grounds when we got back to Dehradun, so I could pester the man in person.

It's ten past nine on Monday morning. I'm waiting for Naha at the front gate. The guard wheels an office chair out into the sun for me. He is dressed in a uniform with a fringed hip sash and a magnificent red plumed beret, as though it were royalty under his protection, not wildlife biologists. Just past the gate there's a small grounds building surrounded by chain-link fencing and topped with concertina wire. A macaque walks casually through the barbed loops.

Naha crosses the institute lawn to retrieve me. As we walk to the main building, he explains that Qureshi is in Monday morning meetings. He walks me to Qureshi's office and promises to let him know I'm waiting.

Qureshi's desk accessories are wildlife-themed — a zebra-stripe pencil cup, a tiger-stripe water bottle. To my left, a sliding glass door opens onto a terrace that, yes, macaques have twice used as an access point, ransacking desktops in a pandemonium of airborne papers and office supplies and then, finding nothing edible, dashing out the way they came. The incidents were related to me, mostly in pantomime, by a man who sits at a desk at the back of Qureshi's office. He speaks no English, and his function is unclear to me. He wears a striped short-sleeved dress shirt and a vest, also striped. It's all stripes all the time in here.

A secretary walks in and places two open file folders on Qureshi's desk. Forms are flagged with Post-it notes. "Otherwise he will sign anywhere!" She laughs. "Does Sir know you're here?"

"Oh, yes. He's in a meeting."

She makes an ominous clucking sound. "His meetings are indefinite. I wish you luck." Behind her, the striped man is tilting into sleep. A macaque minces along the edge of a rooftop at the far side of the courtyard.

Qureshi arrives around 11:00, with a few of his researchers. He is lean and tall and has a warm, sociable demeanor. In place of a perfunctory "How are you," he gives me "How are you finding India? How's your tummy?"

Before delving into the science, we talk more generally about India and its animal predicaments. "The entire country is almost a sanctuary," says Qureshi. He signs as he speaks, working his way through the Post-its. "In the sense that our laws are quite stringent." Since the 1972 passage of the Wildlife (Protection) Act, it has been illegal to kill or capture wild animals without a permit or a state declaration that a particular species is "vermin." Qureshi glances over the top of his reading glasses. "And people are for this."

Hindu deities often take the form of an animal, or they are part animal, or parts of several animals, or their spouse is an animal, or they ride around on one. I share the story of my first visit to Delhi, when a live rat dropped from someplace above the sidewalk and landed on my foot. "You are blessed!" proclaimed the man I was walking with. "The rat is the conveyance of Lord Ganesha."

Qureshi's researchers have been listening intently. "Everything is a deity!" hoots a project fellow named Uddalak Bindhani. "*Basil* is a deity! It is one of Vishnu's wives."

"If you think about it," says Divya Ramesh, a young behavioral ecologist with an easy smile and a pierced eyebrow, "it's a really nice thing, because people have this great association with nature."

But even Hindu tolerance has limits. Especially if the individual is a farmer. India's top agricultural pests also happen to be sacred animals. Elephants represent the deity Ganesh, and monkeys, Hanuman. The wild boar is an avatar of Vishnu. The nilgai, or blue bull, is actually an antelope, but *gai* means cow, and cows are sacred. When state officials wanted to begin culling the animal, they first pushed through a name change. Nilgais are now roj, or "forest antelope."

Despite the endlessly recurring "monkey menace" media froth, the government neither of Agra nor of Delhi has declared macaques vermin. And if they did, they'd be hard-pressed to find exterminators. "You'll not find anyone to kill a monkey," says Nilanjana Bhowmick, a Delhi-based journalist I've recently met. The Veterinary Services department of Delhi's municipal government has a hard enough time finding people to catch macaques even just to relocate them, which is the current strategy. Even non-Hindus avoid the job, as monkey catchers are often harassed and threatened.

Deepening the problem: offerings. Tuesdays and Saturdays, devotees visit Hanuman temples to make a *puja.* To the icons inside, they present coconuts and garlands of marigolds; to the living representatives hanging around outside, samosas and Frooti mango pop. Feeding wild animals, as we know, is the quickest path to conflict. The promise of food motivates normally human-shy animals to take a risk. The risk-taking is rewarded, and the behavior escalates. Shyness becomes fearlessness, and fearlessness becomes aggression. If you don't hand over the food you're carrying, the monkey will grab it. If you try to hold onto it, or push the animal away, Qureshi says, it may slap you. Or bite you. The *Times of India* put the number of monkey bites reported by Delhi hospitals in 2018 at 950.

Qureshi recalls visiting a Hanuman temple in Himachal Pradesh, where he had gone for a wildlife meeting. His hosts warned him not to carry anything valuable, because the macaques would grab it to ransom for food. Qureshi left his phone and wallet locked in the glove compartment of his car. "This one guy came and —" Qureshi stands and pulls his pockets inside out. "Really! They put a hand in your pocket and properly search you!"

I have a macaque story too. Mine is set near Bundi, in Rajasthan, on a trail through the brush to the ruins of a fourteenth-century fort above the city. I knew there were monkeys up there, because their silhouettes are visible along the parapets at dusk. I went in the morning. I had bananas. I was asking for it. I wanted to know what it was like to be mugged by monkeys. My friend Steph followed behind to document the crime with a tightly clutched iPhone. The first shot is of me, looking down, pre-occupied with my footing, orange plastic produce bag dangling from one hand. But look closer. A little tan head has popped up from behind a boulder farther down the trail. Out of the camera's view, another monkey lurks. Outlaws lying in wait for the stagecoach. As I approach the boulder, the first macaque steps into view. While we stand there sizing each other up, the other one shoots out from behind me and grabs the bananas. Slick! I wouldn't call it an attack. It was more of a purse-snatch, over too quickly to create any fear.

In 2008, the Delhi city government passed legislation prohibiting the feeding of wild monkeys, but according to one news story, no fines have been issued. Outside the Hanuman temple in Delhi's Connaught Place, I watched a man approach a group of macaques. There was a side-glancing furtiveness to him, as if he were approaching prostitutes, not monkeys. He quickly handed over a bag of tomatoes and watched as a portly female sat on her haunches, expertly freeing the pulp and sliming the pavement with the skins. A temple employee watched it happen but did nothing.

Because, Qureshi insists, the employee understands the importance of the gesture. "You want to go to heaven? Then you feed them. You want to book a berth, a nice house there? You feed."

"And these are the same people," adds Ramesh, "who are crying, 'Get rid of these monkeys!' "

Qureshi closes a folder and sets down his pen. "Many people, when you interview them, they say, 'Don't kill them!' They just want them to disappear." It's the same everywhere: wildlife NIMBYism. Squirrels in the park are adorable. Squirrels digging in your planters are deplorable.

Qureshi adds that the other problem with government-controlled culling — referring here to the shooting of wild boar and nilgais — is that while it is permissible to kill them, the law forbids eating the meat. "And here" — he means India — "you don't kill a species for the sake of killing. Only a psychopath does that."

The great hope is that science will come up with a way to impose birth control on problem animals. It's true that Qureshi's team has been working on an immunocontraceptive vaccine for macaques, but it's not true that it would render the animal sterile "within minutes," as the *Times of India* reported. Nor would it be orally administered, as other news outlets have reported. Qureshi rests his elbows on the desk. "A birth control pill for monkeys is a farfetched dream." You'd need to make sure enough animals were eating enough of it on a regular basis and somehow prevent other species from doing the same.*

Oral contraceptives are most practical in a controlled, single-species condition. Like a sewer. To control Norway rats, some U.S. cities have begun using an oral contraceptive called Contra-Pest, which relies on two active ingredients. The first, VCD (4-vinylcyclohexene diepoxide), depletes the

---

* Researchers in the UK are gearing up to attempt it, though not, of course, with monkeys. After decades of mass extermination efforts, a National Anti-Grey Squirrel Campaign, and an order of Parliament failed to eliminate "this most undesirable alien" — to quote his lordship the Earl of Mansfield at the June 29, 1937, meeting of the House of Lords — science is stepping up. Researchers have been testing immunocontraceptive bait set out in tree boxes designed to exclude the nation's beloved red squirrels, whose turf the gray immigrants have steadily overrun. So far, as determined by biomarkers in the bait, most of the targeted squirrels in test forests seem to be taking the dose. Good luck, Britannia. Only 120,000 redcoats remain, while the American invaders number in the millions.

ovaries of eggs. VCD began its career as an industrial plasticizer, but when human health and safety tests revealed it to be an endocrine disrupter, it was repurposed for rodent birth control. Because VCD takes time to work, a second compound was added, this one for the boys, too. Triptolide impacts sperm and egg viability for as long as the animals consume it. It's not yet clear that the two together reliably confer permanent sterilization to a community of rats, but a few American cities are giving it a go. It would not, however, seem to be a solution for widely roaming monkeys with plentiful dining options.

The Wildlife Institute of India is running a trial of an injectable immunocontraceptive vaccine called PZP (porcine zona pellucida). The zona pellucida is a protein coat with sperm receptors that surrounds the egg. Give the female a vaccine of foreign (pig, say) zona pellucida and her immune system will be primed to make antibodies against her own zona pellucida. These antibodies glom onto the receptors, leaving sperm with no access to the egg. Fertilization stymied.

Logistical hurdles abound. Like many vaccines, PZP requires booster shots to keep the immune system on alert. This is, of course, a challenge with free-ranging animals. It's time-consuming and costly enough to round up and inject a population of animals the first time around. Administering boosters adds to the time and expense, and on top of that you need some sort of permanent marking — a tattoo, say — to enable booster-givers to know who's had the first shot and who has not.

In the United States, a synthesized zona pellucida has been tried — mainly on populations of geographically confined animals. The wild horses of Assateague Island were good candidates, because they move in herds and it's a small island, so it's relatively easy to give all the injections at one go. And then do it all over again in three to six weeks, after which a yearly booster seems to do the trick. For the tens of thousands of wild rhesus monkeys that roam Indian cities, it makes little sense even to try.

Here's the other problem with zona pellucida vaccines for monkeys. Females that don't become pregnant will quickly cycle into heat again, and each time they do, males will respond with breeding-season behavior. Meaning they're more aggressive more of the time — not only toward other macaques but, it's believed, toward humans.

This happened with white-tailed deer in some U.S. trials of PZP. The bucks weren't aggressive toward humans but they roamed around more, looking for sex, and their wanderings took them across roads and highways and that wasn't good for deer or for drivers. Partly for this reason, immuno-contraceptive research in the United States has focused on a vaccine that instead blocks the effects of sex hormones. GonaCon stops females from cycling. After an initial injection and a single booster shot, 92 percent of mares in North Dakota's horse-dense Theodore Roosevelt National Park remain infertile seven years later. The study is ongoing and the hope is that infertility will prove to be permanent.

Is there any immunocontraceptive vaccine that might confer permanent sterility from a single shot? The National Wildlife Research Center and the U.S. Bureau of Land Management are currently testing one on a subset of a population of wild horses that has outgrown what its rangeland can support. The shot contains two active components (BMP-15 and GDF-9). Antibodies against these components hinder the egg's ability to communicate with the cells that surround and support it, so it never matures. Because this vaccine wouldn't require marking animals and tracking them down for a booster, it would seem to hold promise for treating urban macaques.

Qureshi sees a broader problem with immunocontraception, or any form of monkey contraception, for that matter. That is, people will expect the problem to start going away as soon as the treatment starts. "But you are not killing the animal," he says. "They will live their life." City macaques live twelve to fifteen years. Qureshi estimates it would take seven or eight years before the population drops enough for the effects to become noticeable to the average macaque-aggrieved Indian. "People will say, 'You spent all that money and the problem is not gone?' "

Qureshi politely excuses himself. He has another meeting to go to. Ramesh walks outside with me to hail an autorickshaw and hops in beside me for the short drive to my hotel. Along the way, we pass a dry streambed where people dump garbage. Each time I've passed this spot, pigs or monkeys are picking through the mess. I ask whether there are efforts, as in Colorado, to get the garbage under control.

Ramesh answers that a cleanup campaign is underway. Where before there were only community dump sites, now the city has garbage trucks that drive from building to building. In one locality, the truck plays an inspirational ditty.

I ask how it's going. Ramesh laughs. "The buildings are super multistory, so nobody bothers to go down and put it in the bins. They just drop the bag out the window."

Humans.

One mildly satisfying element of India's "monkey menace" is that the afflicted are often, for once, the upper classes. Urban monkeys prefer parks with trees and other landscaped spaces — the habitats of the well-to-do. From the boughs and greenery, they soon find their way onto roofs and terraces, through open windows, and into the buildings. They raid the mansions and offices of lawyers and judges. Monkeys have turned up in the prime minister's digs and — to the delight of newspaper headline writers — the halls of Parliament.

"They walk around in chambers!" exclaimed Meera Bhatia, a lawyer who has advocated on behalf of the residents of an upscale, monkey-plagued housing development. I met Bhatia for coffee one afternoon in Delhi. She told me she belongs to an exclusive health club, the same one to which Prime Minister Narendra Modi belongs.

"They opened a new pool, and the monkeys were going in the water!"

Bhatia filed public interest litigation on behalf of her housing development, and in 2007, the Delhi High Court decreed that the city's veterinary division must undertake a plan of action. At the moment, the burden weighs heaviest on one man: head veterinarian R. B. S. Tyagi, with whom I have an appointment in half an hour.

Tyagi's office is on the eighteenth floor of the South Delhi Municipal Corporation (SDMC), the bureaucratic heart of Delhi. As if to bestow the proper mindset for your visit, the elevators take ten minutes to arrive. Aggravated civil servants must have jabbed the Up button so often and so peevishly that at some point it gave out. A sign is taped to the wall — DO NOT PRESS REPEATEDLY.

I wait. A janitor pushes a dust mop across the lobby. He walks a perfect line, slowly, almost ceremonially, like a bride coming down the aisle. Another man dust-mops the black marble tile outside the front doors. Say what you will about the government of Delhi, but they've got that dusty floor thing licked.

Tyagi waves me in. He directs me to sit in one of two chairs arranged in front of his desk, even though the other chair is still occupied, the man's business with Tyagi apparently incomplete. Or perhaps he just feels like sitting there a while longer. On the wall behind Tyagi is a framed photograph of a koala. I have barely introduced myself when Tyagi begins talking. "We have been trapping monkeys as per directions of the High Court of Delhi. At present we have two monkey catchers. After trapping we relocate these monkeys to the Asola Bhatti mines. Details in this regard may be obtained from the chief wildlife warden of Delhi government, Dr. Ishwar Singh. Have you met him?"

I have tried. For weeks before I left for India, I would dial the number listed for Ishwar Singh on the forest department website. Day after day, no one picked up. I later came to know that only a sap tries to reach an Indian official through a phone number on a government website.

Tyagi and Singh — the SDMC and the Forest Department — have been trying to pass the buck for years. The SDMC maintains that because monkeys are wild animals, their control is the responsibility of the forest department. The forest department, in turn, contends that monkeys in cities living on handouts are no longer wild, and therefore do not fall within their purview.

I have heard about a plan to surgically sterilize Delhi macaques. I ask Tyagi for details.

"This is again the subject matter of the chief wildlife warden of Delhi government, Mr. Ishwar Singh."

Though I know the answer, I ask Tyagi why no fines have been issued for feeding monkeys at the temples.

"As you know, dealings of the monkey issue are religious. I would like to request you to discuss the issue with Mr. Ishwar Singh, chief wildlife warden . . ."

I finish it for him. ". . . of Delhi government."

"Yes. He will apprise you."

"Feeding is against the law, correct?"

"This is the subject matter of the chief wildlife warden." DO NOT PRESS REPEATEDLY.

While Tyagi and Singh dicker over whose duty it is, Delhi's better-to-do residents take matters into their own hands. Commercial buildings and affluent families hire monkey wallahs, men who patrol with leashed langurs. Indian langurs are the formidable black-faced species I saw in Pauri Garhwal and Rajasthan. They're larger than rhesus macaques, and the macaques keep their distance. A langur sitting on the side of the trail in Bundi was surely the reason I saw no macaques on the way up the hill to the fort. When I looked him in the eye, he raised his upper lip and flashed his canine teeth. Like pushing back a coat to reveal a side-arm, the gesture had the desired effect. I lowered my gaze and moved along.

"The use of langurs is banned." A contribution from the man in the other chair! He introduces himself. He is one of Tyagi's veterinarians. "It is illegal." Because of the Wildlife (Protection) Act. Meera Bhatia told me people in her circle quietly do it anyway. "I don't know whether we can find out *exactly* how many langurs the prime minister's house has, but . . ."* Then she veered off into a story about a macaque that got inside the All India Institute of Medical Sciences and took to pulling IV needles out of patients' arms and sucking the glucose like a child with a straw in a pop bottle.

After a crackdown on the monkey wallahs,

---

* When Donald and Melania Trump visited the Taj Mahal in 2020, the security detail included paramilitary forces, Armed Constabulary units, National Security Guard commandos, and five langurs.

SDMC hired ten men and trained them to impersonate the calls of the langurs. You may go on the internet to see and hear them at work. It is not true, as one newspaper reported, that they dressed in langur costumes. (It appears to be true, however, that the Sardar Vallabhbhai Patel International Airport hired a man to dress in a bear costume to scare langurs off the runways and prevent flight delays.)

"And the impersonators? How did that work out?" I've started addressing my questions directly to the veterinarian in the other chair.

"The problem is not sorted out. At best, the animals will simply move from one place to another. It is not a permanent solution." The same can be said for: Avi-Simian Shock Tape, slingshot brigades, plastic windowsill snakes, langur urine (one man told the *New York Times* that he "had 65 langurs urinating on prominent homes"), and life-size "langur dolls."

Tyagi removes his reading glasses and puts his full attention on me for the first time since I sat down. "Tell me. What is *your* solution to handle this situation in India?" I believe he would truly, desperately love to hear a good suggestion, a fresh idea, something, anything, that might appease his vexatious monkey-loving, monkey-hating public.

I tell him about the success they've had in some U. S. towns, working to lock down the garbage. Even as I'm saying this, I know that in a city as large and chaotic as Delhi it would be a wasted effort. Tyagi withdraws his gaze. "They are not generally going on garbage. That is dogs." Dogs, R. B. S. Tyagi will talk about. Stray dogs are rightly his job. Delhi has more dog bites than monkey bites, and the dogs, unlike the monkeys, carry a respectable threat of rabies. But dog attacks don't sell newspapers the way monkey attacks do. (Nor, for that matter, do the number 1 cause of animal-inflicted deaths in India: snakes. Around forty thousand people die from snakebite every year in India. My Google news alert on *snakes* and *Delhi* has delivered but one hit: a video of a macaque making off with a snake charmer's snake.)

Tyagi replaces his glasses and reaches for a printout on his desk. "This morning I was collecting information about stray dogs in the U.S." He reads aloud: " 'Stray dogs have become one of the most serious public management problems in American cities.' " I think of the cities I know, and the many, many problems that plague them.

"Stray dogs?"

He continues reading. " 'Packs of wild dogs roam American city streets.' Is it true?"

The paper references a story in the *Grand Forks* (North Dakota) *Herald* about a problem on a few Native American reservations. Kind of a leap, that. It makes me wonder: To what extent does the media in India overstate the problem, or even fabricate monkey assault stories? Take the *Times of India*'s "stoning" reference. In a different newspaper's account, a police investigator stated that the deceased had been sleeping beside a pile of bricks that collapsed on top of him when monkeys jumped onto it. Monkey-inflicted death by stone, yes, but hardly a stoning.

I suspect the press lathers up some of these cases, but macaques are certainly compromising quality of life at some of these housing complexes. A keyword search of "monkeys" on India's National Consumer Complaint Forum delivers eight-hundred-some pleas to authorities to, as it is frequently phrased, "kindly do the needful." Here is a typical missive, from Ravi Choudhary of Sector 46, "listing down some of the problems which we face on a daily basis":

1. Breaking of flower pots
2. Breaking of gate, lamp post
3. Children bite
4. Terror in the residents
5. Breaking of electric cables, [fouling of] water tanks Temporarily any langur should be allowed to roam in the sector. Thanks.

I ask Tyagi if the statistic of 950 monkey bites per year is accurate.

"Monkey bites are recorded in the hospitals. That is another department."

Is there any question for which this man has an answer? "Dr. Tyagi, do you like monkeys?"

"Yes, of course. I am a veterinarian."

"I think you like koalas."

Tyagi swivels to gaze at his koala photograph. "Cute animals. *Very* cute. I went to Australia in 2010. I took that photograph." The memories seem to soothe him. He writes something on a pad, tears off the page, folds it in two, and hands it to me. It feels like a kindness, a small melting of the heart. I unfold the paper. It's the mobile number for Dr. Ishwar Singh, chief wildlife warden, Delhi government.

■ ■ ■ ■

Over a span of twelve years, Tyagi's monkey catchers have delivered 21,000 macaques to an abandoned plaster of Paris mine in the southern reaches of Delhi. It was christened the Asola Bhatti Wildlife Sanctuary, and fiberglass walls were erected. There was a plan to plant fruit trees to feed the simian residents, but that didn't happen, so truckloads of food are driven in. I picture barren ground packed with dusty monkeys eating rotting produce.

The monkey area is closed to the public. If you ask, you won't get in, or not without a lot of paperwork or string-pulling. In the autorickshaw that takes me away from the SDMC, I text Nila Bhowmick, the journalist I met: "Feel like a road trip?"

The challenge appeals to her. We decide to just show up and see what we can see.

The staff plays hot potato with us. "You must go across the hall and speak to the forest officer." "You must see Mrs. Prasad in the natural history center." "The roads are closed due to mud." Nila cheerfully keeps at it, walking up to anyone we encounter. Back outside the administrative building, a woman in the parking lot points us in the direction of an elegant-looking man in white: white tunic and breeches, white turban, white mustache looped at the ends like ornamental cursive. He gives his name as Gurji, and he needs a ride out to the plant nursery, which is not far from the monkey area. Just like that, we are in. Nila translates as we drive along.

Gurji works in the nursery, but for three years he had been a macaque caretaker. What Tyagi would not say, Gurji is happy to provide. The city spends 40,000 rupees a day on food for the macaques: apples, corn, cucumbers, cabbage, sprouts, bananas. "Bananas are constant."

It's a six-mile drive to the monkey area. The road is deeply rutted and, yes, in fact, muddy. The sanctuary is enormous, beautiful, and wild, with low mesquite trees and bushy acacias. We pass nilgais and spotted deer. It seems like a fine place to be a monkey, though we have seen none yet. Gurji tells Nila he misses working with them.

"They were your friends," she says.

Gurji laughs. "Monkeys are no one's friends. They'll come and take the food. That's it."

We pull up beside a simple raised platform — a macaque feeding structure. A half dozen of them perch on the edges, eating corn and cabbage.

The sanctuary walls are poorly designed for keeping in monkeys. They climb them with ease and irritate the people who live nearby. With all the food and room to spare, why do they stray? It's the usual problem with translocation. An animal is not translocated into a vacuum. It's set loose in another animal's territory.

"They fight with the new arrivals," Gurji says. "They chase them away. The weak ones go."

We drive Gurji over to the nursery. It's a single-room structure guarded by a langur. It's feeding time, corn and cucumber. The langur ignores the cucumber and ravages the corn, kernels spraying from either side of its mouth. Gurji walks with us to a cluster of simple buildings across the road, once a residence colony for the mine workers and now a small village. A burly male macaque darts up the fiberglass wall meant to keep him out. He has a low brow and glowering mien that make me think of Joaquin Phoenix. He doesn't seem weak. Why would he want to leave the sanctuary?

"They all go in and out," says Gurji. "In my village, you can't even eat a roti without a monkey coming and taking it from you."

He shrugs. "They're monkeys. What can you say?"

Dr. Ishwar Singh answers on the third ring. Asked about the forest department's plans for macaques in Delhi, he replies, "Laparoscopic sterilization!" He says it with grand flourish, as though introducing a special guest. And then he hangs up. It's the shortest interview I've ever done. I call back. I text. I email. No answer. I am the sperm, and he's the immunovaccinated egg. All receptors blocked.

I contact my acquaintances at the Wildlife Institute of India. Project scientist Sanath Muliya provides an interesting answer. He knows of no plans for surgical monkey sterilizations in Delhi, but laparoscopic vasectomies and, for the ladies, tubal ligations, have been underway on and off at eight Monkey Sterilization Centres (MSC) in the states of Himachal Pradesh and Uttarakhand. Since attaining the status of "pest (undeclared)," in 2006, a total of 150,000 macaques have been tied off, stitched up, and tattooed — with an ID number, I'm guessing, or a check mark, or something. Does 150,000 sound like a lot of monkeys? The forest department doesn't think so. In March 2013, the chief conservator for the

Himachal Pradesh Forest Department sent a memo to MSC staff. "It has been noticed that the pace of monkey sterilization has not been satisfactory." The staff are directed to step up the pace "to approximately 90–100 monkeys per day in each MSC."

Photographs of the Himachal Pradesh Monkey Sterilization Centre posted online show two operating tables going at once. Assuming an eight-hour day, that's six surgeries and tattooings per hour that these surgeons need to be doing. Ten minutes per monkey!

But it wasn't the veterinarians who were hobbling the pace. The memo orders forestry officials to triple the ranks of monkey catchers, not medical staff. Even with the gussied-up title of Monkey Capturing Team, no one wants the job. Officials next tried recruiting the public at large, placing a 500-rupee bounty on macaques. Pushback ensued. As an activist pointed out to a BBC reporter, "Monkeys will be captured in a crude manner. . . . Many can get hurt."

Muliya says the public takes issue even with the vasectomies themselves. "They think it is inhumane." Plus, the monkeys scratch their incisions and the stitches come out. "That's what made us pursue PZP as an option," he wrote to me in an email.

Six months after I return home, the *Hindustan Times* would announce a new approach. "To control the monkey menace, the Delhi government's forest department is pinning all its hopes on an injectable contraceptive." But it wasn't PZP, or any other vaccine. It was RISUG (reversible inhibition of sperm under guidance), a gel injected into the vas deferens to block the sperm's path. The article quoted an affidavit "filed by Ishwar Singh, the Delhi forest department's chief wildlife warden . . . before the Delhi High Court . . . [stating] that the RISUG is a viable option after three failed attempts to get NGOs to conduct laparoscopic sterilization." *Laparoscopic sterilization!* The exclamation point was gone now.

The advantage of RISUG, said the *Hindustan Times* piece, is that it is injected. So there would be no stitches to scratch and pull out. But this wasn't true. Sanath sent me the official forest department description of the procedure, which ended with "and the site will be sutured." The advantage of RISUG over vasectomy is that it's reversible. This is of course no advantage with the macaques of India, but it is an advantage for human men (and the women not presently interested in bearing their

171

children). The reason we know that RISUG works on rhesus monkeys is that the human version was recently tested on them.

The woman who'd carried out the tests is Catherine Vande-Voort, division director of reproductive endocrinology and infertility at the California National Primate Research Centers. We spoke by phone, and she confirmed that incisions were necessary. Though I had called to talk about macaques, I was curious about the future of (human) male birth control. A variation of RISUG is headed to clinical trials in the United States. Could people trust it?

VandeVoort thought things looked highly promising. "If you can block a male monkey's fertility, you are golden," she said. Where tens of millions of sperm would constitute a respectable ejaculate for a man, a macaque launches hundreds of millions per wad. "And monkey sperm compared to human sperm look like they're jet-propelled. I've had people who are used to doing human sperm evaluations come into the lab and they look at the monkey stuff* and

* To get the monkey stuff, VandeVoort's team developed a low-intensity penile electroejaculator. Why not just use a vibrator? "Oh, we tried. Oh my god, we tried. You could get a good erection but

172

they're like, *My god, how can you even count them, they move so fast!*" In other words, if it works on a macaque, it ought to work on a human.

However. As a way to reduce a population of wild animals, any form of male contraception comes up short. One busy untreated male can take up a surprising amount of the slack created by his sterilized cohorts. For male sterilization to have a significant impact on a wild population, you'd need to treat 99 percent of the males. Whereas with females, treating around 70 percent should get you there.

Partly for that reason, the sex hormone–suppressor GonaCon is only registered for use on females. Additional reasons are set forth in "Observations on the Use of the

---

they wouldn't ejaculate." They also tried a dummy mount with an artificial vagina. Not happening. "A monkey is not smart enough to understand what we want from him, but too smart to have sex with a dummy." She emphasized that the device does not hurt or burn. Quite the opposite. She told me about an orangutan who'd come running at the sound of her voice. It was memorably uncomfortable: "Having an orang gaze longingly into your eyes as you try to set up your gear."

GNRH Vaccine GonaConTM in Male White-Tailed Deer." When the bucks' testosterone was suppressed, their scrotums shrank, their antlers came in all weird, and they failed to develop "the muscular appearance . . . of mature bucks in rut." Pictured next to a bull-necked control, they're pipsqueaks. Is humiliation an emotion felt by deer? I posed this question to Doug Eckery, who oversees fertility control research projects at the National Wildlife Research Center and serves as its assistant director. "I don't know," he said sensibly.

I asked Muliya how close the state of Himachal Pradesh has come to sterilizing 70 percent of the female macaques in conflict zones. Unknown, he said, because no one has followed up with a population count. I tried to imagine how it's possible to take a census without knocking on doors or mailing out forms. How do you know you're not counting the same individual more than once? Like the anxious San Franciscans whose sightings* led them to believe the

---

* And, likely, what they heard. In a 2017 study by R. Kyle Brewster and colleagues, subjects were asked to listen to recordings of one to four coyotes howling and "yip-yapping" and then guess the number of animals making the sounds. Regardless

174

city's Presidio woodlands held hundreds of coyotes — when in fact it was one wide-ranging pair and their pups. How do people do this?

of background (urban, rural, suburban), subjects overestimated the number of coyotes by a factor of two. Likely leading to "the misperception that coyotes are more abundant than they actually are," and lots of anxious yip-yapping to local and state authorities.

6

MERCURIAL COUGARS: HOW DO YOU COUNT WHAT YOU CAN'T SEE?

For fifty-seven years, California's mountain lions had a price on their heads. The rancher complained because the cats kill livestock, and the hunter, because they prey on deer. And the state listened. From 1906 to 1963, California paid a bounty for each hide or scalp or set of ears brought to a county courthouse or shipped to the Fish and Game Commission. The payments were logged in ballpoint in a series of ruled leatherbound logbooks that now reside in the state archives in Sacramento. On the inside cover of each logbook, someone has penciled the totals for the year, and a foldout map displays the number of payouts county by county. California excelled at counting dead mountain lions.

Counting the live ones has always been trickier. You cannot fly overhead and take photographs, as you might with herds of wildebeests or walruses hauled out on beaches. You could bring in volunteers to tromp through the woods, as you might on a drive count of deer or a line transect of sloths or an Audubon Christmas Bird Count, but it's unlikely any cats would be seen. Cougars are elusive loners. Their presence is knowable mostly by their "sign" — tracks and scat and other unique traces they leave on the land. The most prolific lion hunter in California's history, Jay Bruce, killed more than five hundred cougars in his years on the state payroll, but only once in all that time did he catch sight of one that wasn't under pursuit by his hounds. Up until the 1970s, bounty logs and county kill maps were the closest thing California had to a statewide cougar count. A county with few dead cats was a county with few cats, period.

Ironically, today, if a wildlife agency wants to know how many cougars live within its governance, it may make use of the same expertise that once brought them down. That is why the only state that prohibits the hunting of mountain lions — California — still has houndsmen on its payroll: to count, not to kill. The same skills serve: "cutting" (searching for) sign and, once fresh tracks are found, putting hounds on the scent to tree the cat. What's different is that the dogs will hang back and the lion will be darted and lowered from the tree. Because California seeks not only to estimate the lions' regional numbers but also to assess their genetic health and monitor their habitat use, a handful per region will be GPS-collared and a DNA sample taken.

The Statewide Mountain Lion Project is being undertaken by the California Department of Fish and Wildlife. More specifically, Justin Dellinger. Dellinger's job title sounds like something you'd hear if you asked an animal-besotted ten-year-old, What do you want to be when you grow up? Mountain Lion and Gray Wolf Researcher. That is what it says on his business card. Dellinger has a PhD in wildlife biology, but it was not just his academic standing that landed him the job. It was also his upbringing. His woodsmanship. He grew up in South Carolina, where he divided his time between the woods and his grandparents' horse barn. His hometown is small enough that any time Dellinger was interested in a girl, he had to "go ask Mamaw and Papaw" how he might be related to her. As the first in his family to go to college, he is a source of parental pride, and a little melancholy. He explained to them that he "had to disperse for genetic purposes."

I first met Dellinger at the CDFW Wildlife Investigations Laboratory, where he has a desk that he occasionally and not very joyfully sits at. He wasn't immediately located, so I sat in the waiting area, in the company of the taxidermied specimens I've come to expect at all government offices with *fish* or *wildlife* or *game* or *forest* in their title. A mountain lion crouched, snarling, on a faux rock ledge above the receptionist. A hawk was coming in for a landing beside a rack of hunting information, its talons extended as though reaching for a brochure. Eventually someone brought me back to Dellinger's office, cramped and narrow and rendered more so by a skein of antlers on the linoleum. He finds these and other "l'il wildlife trinkets" at mountain lion kill sites — meaning places where lions have devoured prey. Dellinger is a trophy scavenger, but not a trophy hunter. Killing bucks for their racks is, he said, "not necessarily something I understand." He hunts to be outdoors tracking and to "fill the freezer." As someone who lets slaughterhouse workers kill the birds and animals she eats, I respect that.

At this moment, the meat Dellinger is consuming is pepperoni, on a pizza at the only restaurant in Alturas, California, that is open at 4:30 in the afternoon. A 4:30 dinner is what happens when breakfast happened at 3:30 a.m. and lunch was a granola bar and a tangerine. There is soot on his clothes and face from this morning, when he was following a cougar through a stand of charred pines. It's not that he's had no time to wash up and change. I think it just didn't occur to him to do so.

Though Dellinger is comfortable living out of a tent — is probably happiest living out of a tent — he is by no means a hermit. He doesn't reject civilization; he's just uncharmed by it. His home is less than two hours from San Francisco, but he has never been.

While we wait to order, I ask Dellinger to explain how the counting gets done. I have braced myself for math. The classic animal population estimation method, capture-mark-recapture, relies on proportions and a tolerance for word problems. Say a biologist wants to know the number of woodchucks in a forest. She sets out traps and she Marks, with leg bands, all the woodchucks she catches (M). We'll say she captures 50. She returns them to the forest. A week later, she puts the traps out again and notes how many she Captures this time around (C). Let's say 41. She also notes how many of them have leg bands — that is, how many are Recaptures (R). We'll say 27. She lets them go. She gets out her calculator. Using the formula* M × C over R, and showing her work, she can estimate how many woodchucks live in the forest. In this case, 76. (In a higher-tech version of the technique, it's images of the animals that are captured and recaptured, using motion-triggered camera traps.)

Because the Statewide Mountain Lion Project is a more qualitative survey, Dellinger's method is unique. He calls it "collar and foller," *foller* being *follow* in a South Carolina drawl. Between bites, he describes how the counting part goes.

"It's like, We found this animal, a male, right here." He puts his finger to an imaginary map. The male is given a tracking collar. "Now say we come across another tom track. We pull out the telemetry equipment and check on that first tom. We see that he's nowhere around here, so this must be a different male. Next day we find a female track. Now we've got a minimum count of three." And so on.

* The formula for calculating how much wood a woodchuck could chuck is *0* × (the name is an English corruption of a regional Algonquian word, *wuchak*).

"You don't have to collar any of them, honestly, if you have a few good sign cutters and you use basic reasoning. Let's say you cut a male track over here on this ridgeline." That would be the ridgeline between the oil-and-vinegar caddy and the table's edge. "Now about five ridge lines over" — the back of my chair — "my houndsman finds another tom track, and they're not even close to traveling in the same direction, and they're both from last night. Chances are, those are two different males."

To someone skilled at cutting sign, the absence of it is also telling. Very quickly upon arriving in an area, Dellinger knows whether or not there are lions. If not, he moves on. Still, it will take eight years to cover the whole state. I ask him why more people aren't hired to help. A waitress stops to clear the salad plates.

"Say that one more time? I was distracted." By the busy hand and the remaining pizza slice. "I didn't want to lose it."

One reason the state doesn't hire more trackers is that they don't exist. Aside from the houndsman currently on the job, Dellinger knows of only two others in the state whom he would work with. Both are in their eighties. As for wildlife biologists: "Maybe two percent of them can do this."

I have questions about the "this." How can you tell a mountain lion's gender from its tracks? How do you know where to start looking? How can you be sure a track is fresh from the night before, rather than from the day or the week before?

These things are more easily shown than told, and that is why I too will be eating breakfast at 3:30 a.m.

Tracking wild animals always struck me as something hushed and hidden. I envisioned someone walking with intense focus, head bent, through deep forest. Stopping to examine a broken branch, kneeling beside a watering hole. Moccasins might be worn.

So far it's been louder and less verdant. Dellinger looks for sign while standing at the handlebars of an ATV, driving slowly down a dirt logging road. Driving covers ground faster than walking, and tracks show up well in the fine-grained dirt. (Winter snow provides the other clean canvas for tracks.) We are deep in a forest, but the pines are charred and bare. Twenty thousand acres of Modoc National Forest burned last year, and that is where Dellinger is looking for lions this week.

Where I see blackened trunks in a lifeless moonscape, Dellinger sees the bright green

183

shoots that have sprung up since the fire, thriving in the newly admitted sunlight. Tender new growth is what deer love to eat, and deer are what mountain lions love to eat. Dellinger uses the term "deer specialists." That's what it says on *their* business cards.

To find mountain lions, one heads to the places that would be, as Dellinger puts it, appealing to them. Mountain lions want to be where there is food and water — for them and for their prey — and unstrenuous routes through their territory. A male cougar may walk ten-plus miles in a night of hunting and checking up on his females. A saddle, or pass, between hills or a ridge along the top of them provides an easier, faster way to move through the landscape. Dellinger starts to say that cougars are lazy, then corrects himself. They're *efficient*. A "she" cougar, in particular, can't afford to squander calories. Almost always, a female is either pregnant or has young to feed. A male cub may be as big as or bigger than her by the time it heads off on its own. "Depending on the size of the litter, she may have to kill every day," Dellinger says. "That takes a toll." When he arrives in a new county, he takes out a topographical map and looks for saddles, ridgelines, and draws (small valleys with streambeds).

And national forest logging roads. "A road like this one," he says to me over his shoulder, "that's pretty straight, that gets them from point A to point B pretty quick — they're going to like. And this one is peripheral to the water down there, so it's good hunting for them." Why is there a logging road in a national forest? Because national forests began as — and to some extent remain — managed tree farms for the nation. They were set aside in part, to quote the Organic Administration Act of 1897, "to furnish a continuous supply of timber for the use and necessities of citizens of the United States." (And free grazing space for their cattle. The only animals I've seen in the woods so far this morning have been cows.)

Away to our right, the sun crests a ridge. Bye-bye, moon sliver and Nivea-blue sky. Cue the bird chatter and side-lit glory of a California wilderness dawn. Dellinger is missing it. His mind stays in the dirt. Today he's stopping more than he normally would, to highlight interesting tracks for his back-seat city dweller. He just idled the engine to show me a set of badger tracks. Badgers are un-weaselly members of the weasel family who went their own way, specializing in bullying through ground squirrel burrows to eat the occupants. They evolved long, sturdy digging claws that give their tracks a spindly Edward Scissorhands appearance. To spend a morning cutting sign is to marvel at the surreal variety of feet and dance steps in the animal kingdom. Earlier we saw tracks of a mule deer stotting. The verb *stot* means to spring into the air and land with all four feet at once. (Various theories exist for why deer and antelope stot, as well as several names for the practice, of which *pronking* is the author's favorite.)

Dellinger stops again, this time for the fine-toothed imprints of a ground squirrel's claws. It's been scurrying all over the place, perhaps having caught wind of you-know-who with the Tim Burton manicure. Ground squirrels perform a cousin of stotting, wherein they leap and land with all four feet bunched together, leaving a track that looks like it was made by one large paw. Dellinger says he's watched people track a ground squirrel thinking they were on the trail of a lion. His boss periodically sends "aides" along; most are quietly sent off to cut sign in an area Dellinger knows to have no cougars, lest they drive over or trample tracks they've failed to see. When I first emailed Dellinger asking to tag along, he wrote back, "How are you at tracking?" Attached to the email were two photographs: examples of what I'd be expected to recognize as a mountain lion track in the dirt. A Leatherman multi-tool had been set alongside for scale. Because I could not, at a glance, see any tracks at all, I was puzzled. It looked like a photo shoot for Leatherman.

From what I have learned today, I can now identify deer and badger footprints. I could definitely track a cow. And I can tell coyote and fox from bobcat and cougar. Canid tracks are more elongate and may include claw marks, because unlike cats' claws, dogs' claws don't retract. The tell for a cougar or bobcat track is a pair of prominent notches ("cleats," in tracker lingo) at the rear of the large center pad. These are often visible even if it's a partial track, which, in more gravelly sections of the road, it often is. We brake for cleats.

Dellinger slows to a stop. He tells me to stay on the ATV, which suggests he's found mountain lion tracks and is worried I'll hop off and step on them. He gets onto one knee and lowers his face to the roadway. A tangerine in a cargo pocket bulges like a tumor.

It is in fact a mountain lion. Dellinger lays

a small ruler alongside. Sexing a cougar track is a simple matter of measurement. If it's wider than 48 millimeters, it's a tom. Less indicates a she. And how does one know a she track from the track of a juvenile male? By who else is around. If it's a juvenile, the mother's tracks should be somewhere nearby. This one is a tom.

Dellinger presses the side of his fist into the dirt beside the track for a comparison. "See how this" — the extremely fresh fist-print — "is a lot crisper?" Moist soil holds together better. Taken to the end point, it's the difference between sand-castle sand and hourglass sand. In the heat of summer, by ten or eleven in the morning, the moisture from the previous night's dew will have mostly dried, and tracks from the night before will have lost their sharpness. This is another reason to get out of bed at 3:30 a.m. As is this: the side-arm sun of early morning casts shadows behind the ridges of a track, outlining its contours. If we came back at noon, even Dellinger would miss this track.

Dellinger believes this print to be that of the male he collared near here the week before. The goal today is to find and collar the female who eluded Dellinger yesterday, so we climb back on the four-wheeler and continue to stare at dirt.

Dellinger points to the road ahead of us. "Watch what happens to his tracks as we come up on this curve. See how he's saving steps? He's cutting the corner." If you can read sign, you can infer an animal's intent from the lay of its tracks. If a mountain lion is chasing prey, its tracks are messy and scuffed, and some will overlap. If it's in traveling mode, with a set destination in mind, the tracks are clean and widely spaced. The stride of this lion is shorter. "He's just amblin'," Dellinger says. Looking for someone to eat.

I will admit that, despite a statistically detailed knowledge of the extreme rarity of cougar attacks on humans, these tracks make me nervous — not now, on the back of the ATV, but a half hour later, when I leave the road and go off into the trees to pee. Dellinger never worries. As he put it, "We're not on the menu." Nor does he believe encounters with cougars are on the rise. "Californians are like, 'Lions are everywhere now!' " What's on the rise are home security cameras. Doorbell cameras are the mammograms of wildlife biology. Dellinger slows for a rut. "All it is, is a change in technology." Someone posts a doorbell shot of a cougar. It's reposted, goes viral. News crews show up. The whole neighborhood's talking about it. One sighting turns into five.

Dellinger upshifts for a straight stretch. "They've always been there. We just never saw them." He says he'd bet his paycheck that every day of the year, at least a dozen Californians come within pouncing distance of a mountain lion and never know it.

Almost 10:00 a.m. now. It's warm enough to unzip my jacket despite a steady breeze. Even if we found fresh tracks, the scent would be too dissipated by heat and wind for a hound to track. (The houndsman is off on another logging road; he and Dellinger are in touch by cellphone.) As the sun heats the air, scent molecules become more energetic, bouncing off one another and spreading out, becoming a diffuse cloud of diluted smell. The wind further scatters the scent. Even under ideal smelling circumstances, there will be a point, many points, where a hound loses the scent. This is called, yes, "a lose." To relocate the scent, good hounds will run a zigzag, sweeping wide left and right until they pick it up again. I tried this on a street near my office, after a young man passed by in a reek of Axe body spray. I let him turn the corner and disappear from view, then waited a few minutes. By zigzagging hound dog–style, I was able to track him to his destination, a cheesesteak place on the next block.

Dellinger is ready to pack it in. Tomorrow's another day. Except that it isn't. He neglected to mention that he has to drive to Redding for a "wolf meeting." I picture large canids in business casual. Sensing my disappointment, Dellinger offers to demonstrate how he uses tree-climbing gear to reach a darted mountain lion to lower it down. Without the lion, this is only lightly interesting. He also offers to drive me out to a "community scrape" that he found yesterday. At places where males' territories overlap, or where they make use of the same pass through the hills, say, lions will leave a calling card. Much as the dogs in my neighborhood spritz urine on the same sad shrub in our yard, cougars will kick back the "duff" with their rear paws, depositing scent onto fallen pine needles and other forest floor detritus. Dellinger looks for scrapes under large trees. The bigger the tree, the deeper the duff.

We arrive at the spot and look at some scrapes. While visually unspectacular, a scrape is interesting for what it reveals to those who know the code. When mountain lions scrape, they typically face the direction they're traveling, which is helpful if you're looking for them. And, as with tracks, it's possible to tell how fresh the scrape is — here, by noting how many new needles have fallen on top. Dellinger is explaining how to "sex a scrape." A female kicks back with her rear legs together, leaving two parallel gashes in the duff, whereas a male scrapes with one foot at a time, angling out to the side, "because of its anatomy." By this he means "its balls." Dellinger was recently chastised for incautious language in a press interview. He had, among other things, likened maneuvering a woozy, darted cougar out of a tree to "tryin' to get your drunk friend into a taxi-cab and he's got his hands on either side of the door, resistin'."

I envy people able to read the natural world in this way. I move through the woods the way I flip through Chinese editions of my books, seeing shapes and patterns and having no clue what they might mean. Earlier, Dellinger showed me a line in the dirt from one side of the road to the other, like something made by a kid dragging a stick. It was in fact a "drag mark," but no child passed through here. The line was made by the dangling hoof of a dead fawn being carried in a mountain lion's jaws to a more secluded spot to be eaten. I of course missed the tracks on either side.

When I first met Dellinger, I told him that what I found interesting about the Mountain Lion Project is that it blends modern wildlife biology with its roots in natural history. The early naturalists spent weeks at a time out on the land, tracking and observing, deciphering behaviors, discovering new species. You could sense the excitement in the titles of their journal papers. "Anecdote of a Combat Betwixt Two Hares." "A New Duiker from Zanzibar." I'm sure the authors of "Conservation Phylogenetics of the Asian Box Turtles (Geoemdidae, Cuora): Mitochondrial Introgression, Numts, and Inferences from Multiple Nuclear Loci" felt some excitement, too, but they didn't have those long, glorious stretches alone in the wild.

Dellinger knows the new stuff, but at heart he's a happy throwback. Standing over the skeleton of a deer earlier, Dellinger talked about how he's noticed that mountain lions in less arid parts of the state will pull out a deer's rumen and drag it away before feeding on the carcass. And lions in drier parts of the state don't seem to do this. An ungulate's rumen teems with bacteria that break down the plant matter it consumes; Dellinger's theory is that the behavior slows the rot of the carcass, which might boost the animal's odds of survival in humid climates where meat spoils more quickly.

Naturalists were the original biologists, and hunters and trappers were the original naturalists. No one knew more about a species — the wheres, whens, and whys of its movements through the land and the seasons, its relationships with prey and rivals and mates — than a person whose livelihood depended on that knowledge. The first natural history museums looked very much like the dioramas at a Cabela's. As natural history formalized itself and science became a paying career, rivalries and resentments grew. In 1941, the aforementioned houndsman Jay Bruce wrote a letter to his Division of Fish and Game superior, requesting that he not share Bruce's new report "Cougar in Relation to His Neighbors." "The naturalists have already stolen too much of my discoveries and never give me credit for it," the letter states. "Everthing they should have known, but didn't, they have implied was their own material."

Wildlife biology has always been a kind of snooping. Long before scientists were spying on animals with wildlife cameras or tailing them with radio collars, they were poking around in their excretions. As in human espionage,* it's done because you can't just

---

* The late Sayre Stevens, head of the CIA's Directorate of Science and Technology, had a trap placed in the plumbing beneath the toilet at Blair House, the guest accommodations for visitors of the president, while Soviet premier Nikita Khrushchev was in town. The captured stool was brought to physicians within the CIA's medical intelligence program to see what information it might divulge. Specimens from Egyptian king Farouk and Indonesian president Sukarno (in this case urine from an airplane toilet) were likewise delivered for testing. Given that the tactic predates clinical DNA analysis, how much intelligence could a BM really provide? What could brown do for you? I posed the question to Jonathan D. Clemente, a practicing physician and contributor to the *International Journal of Intelligence and Counterintelligence,* currently at work on a scholarly history of medical support for clandestine operations. Not much, is the answer. "They were looking for blood in the stool, maybe parasites. Whether they got any useful information, I kind of doubt it." More fruitfully, Clemente said, CIA physicians have worked under assumed names at high-profile medical centers where foreign heads of state

ask. You can't inquire of an animal what it eats or how healthy or stressed it is, but you can sometimes learn the answers from its scat.

"Droppings analysis" got rolling in the 1930s. The decade saw a steady progression of learned men prying into the toilet of common woodland creatures: Hamilton on the diet of big brown bats, Murie on coyotes, Dearborn on foxes, minks, and coyotes, Hamilton again on skunks, Errington on badgers and weasels. Before that, if you wanted to know what a species was eating, you opened a few hundred stomachs. You can imagine how amassing enough organs to draw valid conclusions would be an unappealing prospect for most biologists,

---

sometimes travel for care. Why bother with a turd when you have direct access to the man himself and his medical records? Only once, to Clemente's knowledge, has critical information come out of the enemy's toilet. He shared the story of a U.S. Military Liaison Mission that had been spying on a field camp of Russian soldiers. At one point the solders ran out of toilet paper and began using pages from their code pads. MLM operatives went through the soldiers' trash and triumphantly delivered the browned pages to the National Security Agency.

and certainly all the animals. Albert Kenrich Fisher's 1900 "Summary of the Contents of 255 Stomachs of the Screech Owl" made me feel tired and sad, though also vaguely festive, owing to the author's "Twelve Days of Christmas"–style presentation: "91 stomachs contained mice . . . 100 stomachs contained insects . . . 9 stomachs contained crawfish . . . 2 stomachs contained scorpions . . ." Droppings provided a kinder, less taxing alternative.

They still do. Dellinger's master's thesis is on the diets of gray wolves. "I spent a lot of time," he recalled earlier, "walking around looking for poo." (*"Poo"*! There's the California Department of Fish and Wildlife talking again.) Old habits die hard. As we walked around a deer kill site, he bent to pick something up, saying, "Here's some bobcat scat." He held it out to me and then, quickly reassessing, let it drop.

Eventually someone hatched the idea of counting piles of scat to estimate species populations: shit as proxy for shitter. The technique became known as a pellet census: yet more biology-minded humans moving through the wilderness, staring at the ground. Provided your census takers can tell fresh feces from stale, and you know the average number of times a day the species in question defecates, it's possible to work out how many individuals a fecal tally of a set area represents. Possible, but not easy, and probably not all that accurate.

First, your census takers must know their shit. Raccoon scat, for instance, is most reliably distinguished from opossum scat by scent, the latter possessing a foul smell. Ungulates pose a quandary in that they often travel in groups and "dung" while walking. Making it hard to know: Are you looking at the droppings of two individuals or — quoting scatologist Ernest Thompson Seton — one "peripatetic defecator"?

Nor is it a straightforward matter to make the call between fresh and old. No one knew this better than a rat searcher. These were men, "specially trained in the habits of the rat,"* who, among other tasks, boarded ships at England's docks and counted up fresh pellets in order to estimate the vessel's "rattiness." Not as simple as it sounds.

---

* And impressively well dressed for the job. In a 1930 photograph online, a West Indian rat searcher sports a double-breasted jacket with eight brass buttons and the kind of hat more often associated with commercial airline pilots. He carries a spiffy metal box that holds either a rat or a sandwich, I do not know which.

Fresh droppings in a hot engine room could be misleadingly shriveled and dry, and old droppings on a wet deck were misleadingly plump and fresh-looking. Mold was an unreliable indicator, as a 1930 study by Liverpool's assistant port medical officer proved. On certain diets — sunflower seeds, for one, and bran — the pellets grew moldy within twenty-four hours, "whilst on other diets, excreta kept under exactly similar conditions [showed] no signs of mould for several days." And depending on what sorts of edibles were in the cargo hold, it could also be tricky to tell a rat dropping from some other animal's. The small, hard black turds of a rice-fed rat were easily mistaken for those of a mouse. Despite all these complexities, the rat searchers' estimates, when checked against post-fumigation roundups, were impressively accurate, and prompted "a healthy rivalry among the men."

Figuring the daily defecation rate for a species posed its own challenges. Some researchers tried fashioning a "fecal harness" and outfitting a number of representative animals. (The harness held in place the "fecal bag" — a sort of feed bag in reverse.) The results strayed in unanticipated ways. One researcher's harness for shrub-grazing goats proved too restrictive; the creatures were unable to assume their preferred bipedal noshing stance, which allowed them to reach the higher leaves. A subsequent goat-diet researcher published plans for an improved harness that, despite its nineteen leather straps, allowed the goats to rear up on their hind legs. In a minor setback, several of the nonharnessed goats, being goats, ate the leather straps off their pals. Science is never simple.

An alternative would be to spend time spying on animals in the wild. Again, not as straightforward as you might think. "Dunging" frequency, wrote David Welch, in a 1982 study on the "dung-volume method of assessing occupance," differs by time of day and by season. Rabbits in Wales dropped an average of 446 pellets a day in April, when food was plentiful, but only 376 a day in January. The rate also varies by what foods the animal is eating. I'm not just pulling this out of my fecal bag. The Liverpool study details the "enormously" varied fecal output of rats on different diets. A rat fed rice will pass an average of 21 pellets a day; a bran-fed rat averages 128 ("very large, buff-colored cylindrical") pellets a day.

It was around this point that a JSTOR pop-up appeared on my laptop screen.

"Want to connect with leading experts in pellets?" I kind of did. What would they be like? How many could there be? Was I one now?

The future of turd science is bright. Analyzing genetics from scat promises to be a faster, less costly version of capture-mark-recapture. Rather than counting recaptures of marked animals, one would count re-appearances of genetic fingerprints in the collected scat. Dogs trained to whiff out mountain lion scat will soon be brought to these same areas Dellinger has been surveying. Their handlers will bag the scat found by the dogs and bring it back to the CDFW Wildlife Investigations Laboratory for genetic analyses. Scat can also deliver information about the health and genetic diversity of cougars in different regions.

If the genetics work turns up population data similar to what Dellinger is getting from his tracking and collaring work, it means scat-detection dogs and gene sequencing can be trusted to do the work on future surveys. If all goes as hoped, Justin Dellinger will be replaced by a heap of shit.

I think he'd miss being out here, though he says no. He says it'll free him up to spend more time researching hazing techniques and other ways to keep mountain lions apart from humans. Because whenever the two cross paths, at least in California, controversy boils. As Dellinger puts it, "For some people, ten is too many. For some, ten thousand is not enough." Interestingly, it is not commercial ranches requesting the majority of California's mountain lion depredation (kill) permits. Between 70 and 90 percent are granted to backyard famers — people with between two and ten animals. (The state has few large commercial ranches.) From a distance, the killing of a cougar is, for many in my state, an affront. *Build a safe nighttime enclosure for your animals! Keep your pet inside at night! How is the life of a beagle or goat worth more than the life of a wild mountain lion?* A bellows to the flames is the common impression, based on heavy media coverage of a pocket of cougars isolated by the Los Angeles freeway system, that the species is imperiled throughout the state. California's cougars are neither endangered nor threatened. But they are big and they are beautiful, and those are the animals people fight for hardest. It's the everlasting politics of charismatic megafauna.

And megaflora, even. The bigger the tree, the deeper the duff. Or something.

7
WHEN THE WOOD COMES DOWN: BEWARE THE "DANGER TREE"

What a Douglas fir does, it does very slowly, and that includes dying. Possibly the least attractive feature of a nine-hundred-year life span is the century or two spent dying. Decomposition drags on for another hundred years or so. A tree is the rare organism to which the comparative *deader* is often and accurately applied. A recently dead, or "dead hard," conifer progresses to "dead spongy," then "dead soft," limbs and top rotting and dropping off, until the last piece of standing trunk topples and the tree enters the final classification, "dead fallen." At some point in its protracted twilight, a tree that stands near a road or path or building may earn a new classification: "danger tree." Because if it falls, anyone it lands on will spend a very, very short time dying.

The victims of arboreal manslaughter may be, unlike the perpetrators, quite young. The *Australian Journal of Outdoor Education* published a summary of cases, since 1960, of children (and in two instances, their teachers) killed by falling branches or trees during school camping trips: six killed sleeping in their tents, one while swimming near a eucalyptus grove, and another six killed while hiking, including two teenagers crushed when the top of a mountain ash tree broke off and rolled down a hill onto the trail.

Wind is a common accomplice. The journal *Natural Hazards* reports that in the United States, between 1995 and 2007, trees toppled by strong winds caused the deaths of nearly four hundred people. My husband, Ed, and I were twenty feet away from a similar fate, awakened early one windy morning by the crack of a large branch breaking off an oak and landing near our tent.

Some trees kill in the normal course of life. The Coulter pine drops a cone as heavy as a bowling ball. According to "the largest review of coconut-palm related injuries," sixteen Solomon Islanders were struck by falling coconuts between 1994 and 1999. Balinese newspapers have three times in recent years reported on cases of bodies found beneath durian trees. The fruit of the durian tree makes an excellent murder weapon: big, heavy, and covered in hard spikes. The "suspect," being a tree, cannot hide the weapon; a bloodied fruit lay beside one victim's head. It is difficult for authorities to generate caution or concern. Confronted with the sign "Dropping Pine Cones, Proceed at your own risk," most will proceed.

The term "danger tree" is itself somewhat hilarious. It's like "danger mitten." The staff of Vancouver Island's MacMillian Provincial Park, home to a stand of "legacy" conifers, find no humor here. Because the most geriatric trees are also the tallest and most majestic. They are the trees people pay money to hike and drive amid and the ones the public very badly doesn't want cut down. This creates a conundrum and, very occasionally, a tragedy.

In 2003, an Alberta couple were passing through MacMillan's Cathedral Grove, a stand of massive, centuries-old conifers, when a fierce snowstorm hit. They pulled off the roadway to wait it out. One of the ancient firs, overburdened by snow and weakened from rot, fell onto their car and killed them.

Since then, MacMillan has maintained a relationship with a certified danger-tree assessor. Twice a year and after any big storm, ongoing for fifteen years, Dean McGeough roams the woods looking for signs of dangerous decrepitude. Today is one of the semiannual inspections. Over the course of the day, Dean will flag trees he deems in need of mitigation: a limb lopped, or a top, or something more drastic. This is the part where, statistically, most of the manslaughter goes down. By a large margin, the people whom trees kill most often are the people bringing them — or pieces of them — down. An on-the-job fatality for a faller, as chainsaw wielders are known in these parts, is sixty-five times more likely than it is for workers in general. These are men with compression bandages stuffed in their pockets the way my grandmother had Kleenex. Men whose fabric of choice is cotton-Kevlar. Though it isn't usually the blade that kills. It's the tree. Sometimes it's the one they're cutting, but more often it's a bystander. The tree may bend the branch of a neighbor as it falls, causing it to slingshot back at deathly speed. Pieces of other trees lodged in the branches — a "swinging snag" or an "insecure hang-up" — may pull free and come down on the faller.

British Columbia has an active Forest Safety Council, and two members are here today. The title given on their business cards is Falling Safety Advisor. Earlier today I met a Falling Supervisor. The word *falling* has lost its slapstick for people in the logging industry. Someone will mention a guy from a long-ago job, and someone else will go, "Is he still falling?"

The most dangerous trees to fall are (duh) danger trees. A healthy tree with sound wood can be made to fall in any direction. Like this: Rather than cutting straight through the trunk, the faller stops partway and goes around to the opposite side and makes a sloping undercut. Now when he goes to finish the cut, the trunk will tilt down onto the slope made by the undercut and fall in that direction. A rotting tree is hard to control in this manner, its fall impossible to predict with certainty. If a conifer is rotting from the top down, the softened portion may break off as the tree starts to lean and come down on the faller. Or the whole rotted trunk may "telescope" — collapse straight down into itself. Or a rotted portion of trunk may suddenly crumble and change the direction of the fall. Think of those osteoporotic olds whose bone has grown so porous that one day a hip gives way when they shift their weight. (All these "overmature" trees may explain why the lumber company that once owned the grove donated it to the province in the first place: lotta punky lumber.)

Ideally, no one should be anywhere near a danger tree when the wood comes down. That is why very tall, very old, very dangerous trees are not cut down but, rather, blown up. Explosives aren't exactly crib toys, but they can be detonated from a safe distance. So regardless of what comes down and from what direction, no faller will be felled.

After Dean finishes his inspections, expert faller blaster Dave "Dazy" Weymer will start work. (The nickname dates to his twenties and has to do with weed, not flowers.) Dazy has been blasting trees for thirty-five of his sixty-eight years. Both his father and his grandfather were loggers. He grew up in logging camps. He was, he says, "a bit doomed to be a logger." I first saw Dazy in a YouTube montage, a sort of highlight reel of explosions and screaming chainsaws against a booming soundtrack of insistent strings and kettle drums. You need ear protection just to watch his videos.

The forest floor in Cathedral Grove is hardly like a floor. It's an obstacle course of decomposing branches and logs, their surfaces and outlines obscured by a damp, spongy pelt of mosses and ferns. It is difficult to predict when your foot will connect and what will happen when it does. It may come to rest on a log or it may push straight through what appears to be a log but is in fact crumbling, log-shaped mush. You will stumble and fall, but you won't be hurt, just moist. Moist fallen.

While Dean makes his rounds, he and Dazy bring me up to speed on basic tree anatomy. The tree, I am learning, is not entirely unlike the human. The older, harder wood that runs through the tree's core serves as the skeleton that supports it. Surrounding this spine of "heartwood" is the "sapwood," the flesh through which courses, slowly, so slowly as to possibly demand a different verb, the blood of the tree — the sap.

Bark is of course the tree's skin. It protects the flesh, and it is — again, like our skin — both an entry point for infection and a part of the immune system. Conifer bark secretes resin (aka pitch), a thick, sticky goo that seals wounds, traps bark beetles, kills pathogens. Also like us: a tree's crown thins as it ages, and the point at which its circumference is greatest is called the butt. And there my trees-as-people comparison sputters out.

"There's one I blasted." Dazy has a deep voice that projects well when he needs it to, which he often does because he's holding a conversation across a grove of trees or over the sound of an idling chainsaw. He's pointing to a Douglas fir. These trees stand out from the others here by their bark — thick, with deep vertical rifts.

From ten feet away, looking straight at it, the blasted fir looks no different from the live intact legacy trees all around it. Only the top third is gone, and to see the top third of this formerly 180-foot tall tree, you would need to crane your neck all the way back. Removing just the upper third makes the tree lighter and more stable — less dangerous — and at the same time preserves the grove's medieval Sherwood Forest vibe, what the tourism professionals call "visitor attractiveness." At eye level, the living, the dead, and the blasted look the same: enormous mossy tree trunks. As Dazy says, "You wouldn't know it wasn't just another pretty tree."

Elderly trees perform their own, more subtle version of what Dazy does. It's called retrenchment. The tree's trunk circumference and roots continue to grow, but it stops getting taller and the limbs of the crown die back and drop off. This makes it less top-heavy. More importantly, there is less "sail," meaning fewer surfaces for the wind to catch and less blowing about of the crown and risk of what forestry people call "wind-throw" — a tree uprooted and blown over by powerful gusts.

I lean way back to try to see Dazy's handiwork on the fir. This causes me to lose my footing and topple backward off a log. Falling Author. Dazy extends a hand. It is noticeably unlined for his age, probably because when he's outdoors he's usually wearing gloves. If other fallers read this, he will no doubt get grief about his lovely hands, but I believe a man named Dazy will handle it.

Here is another reason not to cut a danger tree down to a stump. Dying and decomposing trees, far more so than young living trees, provide real estate for wildlife. Rot-hollowed trunks become dens for bears. Dead tree branches are hunting perches for raptors. Soft, rotting sapwood is easily excavated by woodpeckers and other cavity nesters. For this reason, a "danger tree" is often classed as a "wildlife tree," too. Blasting off the upper third of the tree facilitates the process. It speeds the decay of the remaining trunk by letting rainwater seep into its interior via the jagged, open terminus — the point where the blast took place. Dazy holds back a branch for me. "Biologists love blasted tops," he says as I step through the opening. Provided, that is, that the work is not done during anyone's nesting season.

Dean has marked a large Douglas fir for action. He breaks off a leathery-looking disk, one of a half dozen protruding from the bark. "This is a conk," he says, handing it to me, smiling vaguely. Dean has a kind of ongoing low-grade smile, though never really seems pleased. A conk is the tip of the iceberg, rot-wise. The symptoms of fungal infestations are often hidden until the disease is well-entrenched. By the time conks show up on the outside of a tree, the inside is far rotten.

Still, there's no real rush to take action. This tree has had conks the entire fifteen years Dean has been monitoring it. The bark comes off easily in vertical chunks, like wax drippings off the side of a candle. Dean breaks off a piece of bark and crumbles it between his fingers. Insects take advantage of the punkiness to work their way in and lay eggs. Contributing to yet more punkiness. "See this white powder?" Dean says. "This is frass." Frass is insect excreta, and my favorite new word of the day. It replaces *kerf,* which refers to the width of the space left by a saw-blade cut and is a useful word for Scrabble.

Dean walks out to the tree's drip line, the outermost reach of the branches overhead. This typically indicates the end point, underground, of a tree's roots. He shows me where the root mass is starting to lift on one side, because the tree is leaning. Danger tree! Dean adds it to the work list for tomorrow.

Lately, the trees of Cathedral Grove are succumbing to a root rot called *Armillaria* that spreads underground, an infected tree passing the fungus to its neighbors where the roots touch. Cedars prevail. They have chemicals that resist many of the fungal rotters (thus the wood's popularity for roofing shingles and outdoor furniture). The current situation in the grove is ideal for cedars. They need a fair amount of light, so as their less rot-resistant neighbors perish and fall, the cedars benefit from the newly admitted sun. It's all cyclical, Dean is saying. At some point drought will take the cedars, and another species will thrive where they've perished.

Dean does a lot of tapping and sounding to gauge the extent of a tree's inner rot. He's been dropping Latin names faster than

I can misspell them. Dazy keeps it simple: heart rot, butt rot, root rot. Dean and Dazy used to teach a falling safety class together. Dazy would talk about technique, and Dean covered regulatory matters. Dazy would throw some *fuck*s into his first lecture to put the students at ease. Dean is not a cusser. His gear is immaculate, and he fills out paperwork promptly and legibly. He is exactly who you want keeping records on dozens of two-ton trees that might topple over onto people.

As different as these two men are, they are similar in the degree to which they don't fit my lumberjack stereotype. A few minutes ago, the group was comparing notes on their diets. Dean has two friends who each lost forty pounds on the keto diet, eating bacon "like it's going out of style."

"Man, I could do that," a falling safety advisor said dreamily.

Dazy volunteered that he is doing high-fat/low-carb, for his heart, but remains wary of bacon. "I try to sorta dwell on avocados. And fish."

"Fish," Dean agreed. "There you go."

Dean has tagged six trees for blasting tomorrow morning. We set a time to meet back here and call it a day. No one goes for beers. Carbs and all that.

The explosives are stored in an unmarked silver shed in the woods, five miles up a dirt logging road. *Shed* is the wrong word. Technically, an explosives storage structure is a "magazine."* This one has walls six inches thick and filled with gravel, so yahoos and hunters with poor aim can't shoot through them and blow the surrounding forest to mulch.

It's 5:00 a.m. The sky is still black, the Milky Way at maximum milk. A half dozen men from a road crew mill around in the headlights of trucks, carrying bags of Austin Powder Company explosives. I watch Dazy load five "sticks" of Red-D into the bed of his truck. It comes in plastic tubes and looks more like cookie dough than dynamite. Like many products in Canada, Austin's are bilingually labeled. "*Explosifs,* Explosives" — a rare instance where the French is

---

* You can learn all about these structures on www .explosivestoragemagazine.com, which I at first took to be an online periodical about explosives storage. But it's just a redundancy. The industry does have its own periodicals, however. For instance, the *Journal of Explosive Engineers,* which I would subscribe to for the title alone.

briefer. At a supermarket in town I saw a bag of *"nourriture pour oiseaux sauvages."* Birdseed. Dazy wires a Day-Glo sign to the cab of his truck: TRANSPORTING DANGEROUS GOODS. Now if we're in a fiery crash on the way to the grove, the emergency responders will know to keep their distance. When we arrive, a morning meeting is convening around the hood of someone's truck. Dean is here, and the falling safety advisors, and some men to cut up, or "buck," downed treetops. Because the trees are near the highway, coners and flaggers are also here, to halt and direct lanes of traffic.

Dazy steps into his climbing harness and gets ready to ascend the first tree, a fir. He buckles climbing spurs onto his lower legs. By kicking the spurs into the sides of the tree — left, right, left — he ascends the trunk. Holding his upper body is a flip line looped around the trunk and into his harness. After every few kick-steps, he uses the flip line to pull his body in close to the trunk, then flips the now slack line a foot or so higher. And repeat, all the way up to the height at which he'll bore the hole for the explosive. Dazy has no fear of heights and has never fallen. "That seems to me a once-in-a-lifetime sorta maneuver," he said, when

I'd asked.

It's cold and drizzling and barely light out. A safety advisor lends me a work coat. Wood chips in the pockets. I can hear the flaggers' chitchat over Dean's radio handset. They're at either end of the work zone, alternately stopping and waving on single lanes of traffic. "Hey," one radios to the other. "Here comes yer girlfriend."

Dazy lets down a rope, and the falling safety advisor ties the chainsaw to it. "There's probably a special knot, but we're not going to use it."

The chainsaw ascends and Dazy unties it and then lets us know the rope is coming back down. Sawdust and noise begin spewing from his perch. When the hole is bored, the chainsaw Rapunzels down, and a backpack with the explosives goes up.

Fifteen minutes later, Dazy's work is done. He climbs back down, trailing fuse. Dean spools it out to the detonation site, three hundred feet distant. We all follow. The flaggers radio that traffic is stopped in both directions, and Dean blows the air horn, twelve blasts. I'm the guest detonator. I get to stomp the "thumper." This sets off a chain of minute explosions that travel, in an instant, along the shock tube fuse. Now comes the boom, followed by two sharp cracks as the tree's top crashes through branches of an adjacent tree, then the thunderous *whump* as it hits the ground and a coda of exuberant whooping from everyone except Dean. If a tree falls in the forest and no one is there to hear it, that's a shame.

Dazy leads us back to the blast site. The accomplishments of "fragmentation and heave" lie all around us. A team of buckers slice the fallen top. The remaining tree, from our viewpoint down here among the mosses and ferns, looks just the same. Though of course it is different. It's safer.

The safety advisor is still grinning. I am, too. I'm not sure why big (controlled) explosions cause humans such glee. We seem to be drawn to extremes: huge, tall, loud. It's the pull of awe. It's one reason we care about whales and not sprat,* why people hug trees and step on clover.

No surprise, then, that Dazy's work in this grove has from time to time drawn complaint. He once tried to talk to a protester, to explain that these trees were dying, and

* The Abrau sprat is among 455 critically endangered fish, none of which are featured in conservation fundraising campaigns. Who will save the eightgill hagfish? Who cares about the razorback sucker and the delta smelt?

that they'd be coming down soon(ish) anyway. To which the protester replied, "We think the trees know when it's their time to fall down." Of course, it is not knowledge that prompts a tree's fall, but some fatal brew of wind and gravity and damage and rot.

I can't judge. We all have emotional connections to certain branches of the tree of life, and for some that branch is trees. We are irrational in our species-specific devotions. I know a man who won't eat octopus because of its intelligence. Yet he eats pork and buys glue traps for rats, though rats and pigs are highly intelligent, likely more intelligent — I'm guessing, for I have not seen the SAT scores — than octopuses. Why, for that matter, is intelligence the scale by which we decide whom to spare? Or size? Have the simple and the small less right to live?

Trees, the elders in particular, seem to evoke an urge to protect and defend. Perhaps that's because the trees can't do it themselves — or not in ways easily evident. A tree can't run away or fight back against anything larger than a beetle. Trees are vulnerable, peaceable, innocent. Plants in general have that vibe. Don't be fooled.

8

THE TERROR BEANS: THE LEGUME AS ACCOMPLICE TO MURDER

Like the FBI, the USDA has its lists of top criminals. Noodling around on the Federal Noxious Weeds List and other roundups of most wanted invasives, I came upon a plant called the rosary pea (or in India, the jequirity bean): *Abrus precatorius.* What caught my eye was a photograph of the plant's seed, a striking red and black bean instantly familiar to me, because I have two of them on my desk at home. They were given to me on a rain forest walk in Trinidad, by a guide who called them jumbie beads and said that locals wear them to ward off evil spirits. What he didn't say, and maybe didn't know, was that the pretty seeds of *Abrus precatorius* are the source of abrin, arguably the most lethal phytotoxin (plant toxin) on earth. Abrin is on the U.S. Health and Human Services list of Select Agents and Toxins, alongside the likes of ricin and Ebola virus. Possession of anything over a gram of abrin is a federal crime.

Possession of rosary peas, however, is legal. The internet has thousands of rosary pea necklaces and bracelets★ on offer, as well as crafting websites selling the beans in bulk to people who make these items. I looked at my beans and I thought of all the times the grandkids had been over. What would happen to a toddler who picked them up and swallowed them?

Likely nothing much. The seed's hard casing stands up to gastric juices and travels the gut intact. Fortunately, toddlers don't have molars during the "oral exploration" phase, when they try to put the things they encounter into their mouth. A parent whose child got into someone's rosary peas might only know about it because the toddler had begun shitting jewelry supplies.

Virginia Roxas-Duncan is a supervisory biologist with U.S. Army Medical Research Institute of Infectious Diseases, which researches biowarfare countermeasures. She wrote about abrin for the *Journal of Bioterrorism & Biodefense* and played with rosary peas as a child in the Philippines. "He had

★ But oddly, no rosary-pea rosaries. And only one 1931 JEQUIRITY BEAN PURSE ★POISON IF EATEN★ — as the seller listed it.

diarrhea," she said, of a playmate who once ate some. "But the following day we were playing again."

Even individuals who chew their rosary peas will likely pull through fine. Attempted suicide by *Abrus precatorius* is not uncommon in rural southern India, where the plant is easy to find and other agents of self-death are not. In 2017, the *Indian Journal of Critical Care Medicine* published a review of 112 suicide attempts. Six ended in death. In 14 percent of the cases, there were no symptoms at all.

It's much the same story with castor beans, the source of abrin's higher-profile cousin ricin. Like rosary peas, they are easy to come by legally, as both the plants and the seeds are sold by nurseries as ornamentals. (Though if you clean out their entire stock, as one questionably stable Washington State individual did, staff may alert the FBI.) *Clinical Toxicology* reviewed cases of castor seed ingestions logged with a Midwest poison control center over a ten-year span, 84 cases in total. Forty percent were suicide attempts — using a median of 10 castor beans. The other 60 percent were unintentional ingestions with a median of 1 bean — likely those intrepid diapered oral explorers. The seeds were crushed or chewed in 60 percent of the cases. No one died or was seriously ill as a result. Mostly there was vomiting and diarrhea.

Oddly, swallowing pure ricin, rather than castor seeds, appears (from mouse data) to be even less likely to be fatal. Montana State University biochemist Seth Pincus studies the toxin's therapeutic potential, and is developing treatments for people exposed to it. In his own tests, mice consumed the equivalent, for you or me, of about a Coke bottle's worth of concentrated ricin before succumbing. Pincus's theory is that oral bacteria may be absorbing the pure toxin, and stomach acid and enzymes degrading the remainder. Whereas if someone consumes ground castor seeds, the plant material acts as a sort of time-release mechanism, protecting the ricin in the mouth and stomach and delivering it fully operational to the intestine.

Just by the way, it is not ricin that makes castor oil an effective purgative. As the International Castor Oil Association takes pains to assure us on its website, the ricin is left behind when the oil is extracted from the seeds. Unless death by diarrhea* and

* A technique favored by Mussolini's squadristi thugs. Political foes were force-fed large quantities

dehydration is the intent, castor oil is a useless murder weapon. Casey Cutler had not visited the International Castor Oil Association website before he headed out to an Arizona Albertson's in the summer of 2005 to buy castor oil for the purpose of extracting ricin. George Smith, a senior fellow with GlobalSecurity.org, details the case on theregister.com. Cutler owed money to a drug dealer and had hatched a plan to offer ricin masquerading as recreational drugs, should the man show up to collect. While Cutler was tinkering with his castor oil, his roommate began feeling ill. Fearing it might be ricin poisoning, the roommate went to the emergency room. It was just flu, but at the mention of ricin, medical personnel called in a potential terrorist situation and a Phoenix SWAT team descended upon the apartment. Cutler served three years for, essentially, possession of a laxative with criminal intent.

Cutler had one thing right: ricin via a

---

of castor oil — up to a quart, according to The Straight Dope. *Who does that?* Moreover, why? To kill by dehydration? To humiliate? I could find no satisfying answer, not even from the International Castor Oil Association, which, despite large quantities of emails, had no comment.

needle in the arm would — if you actually had some in the needle — most certainly kill a man. By injection, the lethal dose (for a mouse) is in the neighborhood of a millionth of a gram. In 1978, Bulgarian dissident Georgi Markov was assassinated by a speck of ricin shot into his thigh with a pneumatic spy umbrella as he stood at a crowded London bus stop.

Assassination by abrin injection dates at least as far back as the nineteenth century, when a spate of cow killings was linked to a group of leather workers in the south of India. The technique is laid out in detail in *Pharmacographia Indica: A History of the Principal Drugs of Vegetable Origin, Met With in British India.* Ground rosary peas were made into a paste and shaped into a stout needle, called a *sutari.* This was then dried in the sun, honed, and affixed to a wand. As the cow was whacked,* the point would

---

* A few *sutari* crafters branched out into murder for hire. An 1890 issue of India's *Police Gazette* profiles the "great poisoner" Dooly Chamár, eventually caught and sentenced to "transportation for life." As a punishment, a lifetime of riding public transportation in India made a certain amount of sense; however, that's not what was meant. It meant he was exiled. Also initially

225

break off under the skin, leaving little trace of the crime.

And now I understand why intercepted terrorist communications have, on occasion, mentioned plans to explode suicide bombs that contain ricin or abrin. The shrapnel would act as tiny *sutaris,*★ injecting the poison into otherwise survivable flesh wounds. "To give the conventional bomb a more lethal effect," explained a piece in the online edition of *The Diplomat.* Whereupon the paragraph broke for a subscription come-on: "Enjoying this article? . . ." I don't

---

confusing was the locale of exile — a white-sand idyll in the Andaman and Nicobar Islands. Before the dawn of tourism, the islands housed a penal settlement, where British colonials tested torture techniques far more heinous than anything dished out by Indian Railways.

★ The militaries of various Western nations have experimented with this from time to time, but they did not call them *sutaris*. They called them "flechettes" — tiny toxin-dipped arrows, as many as 35,000, packed into a bomb. Both Canada and Great Britain tested flechettes on animals. Despite impressive efficacy, neither nation, perhaps recognizing the rather dim threat of a projectile that sounds like a feminine hygiene product, added them to the arsenal.

know, I had to say, does one *enjoy* an article about the mass slaughter of innocents via poison-tinged shrapnel?

You may be thinking, Did these terrorists perhaps intend for innocents to inhale the toxins? Perhaps so. Ricin administered in this manner is approximately as lethal as it is by injection. "Massive pulmonary edema," Pincus volunteered. "You drown in your body fluids." However, to kill a crowd of people in this manner, your terrorists would have to possess equipment and expertise sufficient to create a cloud of extremely fine ricin aerosols — ideally no larger than one or two microns. Otherwise the mist won't stay airborne long enough to pose a threat to large numbers of people. (Fine aerosols, compared to droplets, also penetrate more deeply — that is to say more dangerously — into the lungs, a discovery I was very much not, in the midst of a COVID-19 pandemic, *enjoying.*) Anyway, the terrorist cells in question — al Qaeda in the Arabian Peninsula (ricin) and Jamaah Ansharud Daulah (abrin) — did not have sophisticated aerosol dispersal systems. They had humans wearing bombs, bombs more likely to incinerate a toxin than to disperse it.

It's also possible that these groups planned

to add ricin or abrin to their bombs for the simple purpose of terror. Being terrorists and all. Because regardless of how many people a ricin incident harms, it seeds fear in a million more.

"Know the facts. Protect yourself." So go the ominous, ever-present phrasings of public awareness campaigns. You see it on websites about HIV, dengue fever, and Zika virus. Lead poisoning, identity theft, date rape, TOXIC BEANS.

The capital letters are not mine. The words appear that way on a Utah food-handler testing and certification site. The threat they refer to is not jequirity beans or castor beans. (The castor bean isn't actually a legume. It's a spurge.) The threats are kidney beans, red or white, broad beans, and lima beans. Fail to boil these common edibles for at least ten minutes, and you may find yourself in significant gastrointestinal distress. As did a thousand-plus viewers of a Japanese TV show that recommended grinding white kidney beans in a coffee mill, toasting for three minutes, and sprinkling on rice. According to the journal article "The 'White Kidney Bean Incident' in Japan," a hundred people were hospitalized.

For more on the evil a kidney bean can

inflict upon a human, I steer you to "Foreign Body (Kidney Beans) in Urinary Bladder: An Unusual Case Report." In 2018, in Jaipur, India, a young man pushed four kidney beans up his urethra "for the purpose of sexual gratification." As often happens in these cases, the items made their way beyond the point of easy recovery, and once discomfort outweighed embarrassment, the man sought medical help. An ultrasound revealed the kidney beans "floating" inside the man's bladder. Because the beans had been soaking overnight, they had, as any dry bean would, expanded and softened, complicating their extraction. Figure 3, "kidney beans removed piecemeal" is a photograph: a helping of broken, slightly mashed beans in a stainless steel surgical basin — more appetizing than most things removed by forceps in a surgery suite, but probably not, owing to the urine presoak, more delicious.

Is there something uniquely dangerous about beans? I posed this question to plant scientist Ann Filmer, recently retired from the University of California, Davis. In her reply, she included a link for a website she had put together on poisonous garden plants. I was taken aback to note that nine of the 112 plants in Category 1 (Major

Toxicity: "may cause serious illness or death") were currently, or had recently been, growing in our yard: oleander, lantana, night-blooming jasmine, lobelia, rhododendron, azalea, toyon, pittosporum, and hellebore. Another, the houseplant croton, was growing in an orange ceramic pot in my office.

In other words, it's not beans. It's plants, period. If you can't flee or maul or fire a gun, evolution may help you out with other, quieter ways to avoid being eaten. Over the millennia, natural selection favors eaters who turn up their proboscis at you, and eventually they all steer clear.

Given the surprisingly large number of deadly garden plants, why is it that ricin gets all the press? Why don't terrorists and assassins extract toxins from these other plants? The answer to the second question likely resides in the first: ricin gets all the press. The Markov murder made ricin a bright, blinking thing on the terrorist radar. It became the go-to poison for two-bit killers and survivalist crackpots. You don't have to go to the dark web to find instructions for extracting ricin from castor seeds. A quick Google search will take you there. But unless you're a criminal *and* a chemist — like Walter White, who extracts a toxin from lilies of the valley in the fourth season of *Breaking Bad* — you probably lack the equipment and know-how to turn any of these other plants into an accomplice for murder.

Ricin's notoriety has given it a sinister cachet other phytotoxins lack. If you're trying to build some cred in terrorist circles, it sounds better to say you're making ricin than to say you're trying to extract something from a rhododendron. This was pointed out to me by Andy Karam, a counterterrorism professional and the author of *Radiological and Nuclear Terrorism.*

And yet. None of this explains why ricin and abrin are the lone plant poisons on the HHS Select Agents and Toxins list. Seth Pincus had an answer for that. Ricin and abrin, he explained to me when we spoke, are "promiscuous" toxins. Ricin makes its trouble by binding to galactose, a carbohydrate on the surface of living cells of all types. (The outermost layer of skin cells is dead, so touching ricin powder would not pose a danger. Save a stamp, would-be postal assassins.) Most other deadly toxins — cholera toxin, say, or botulinum toxin — wreak their havoc exclusively in one site: cells of the colon, say, or on nerve cells.

I forwarded Pincus a web page from a

Chinese chemical supply vendor. "RICIN in stock with best price," it said. (I'll say: $150 for a kilogram of ricin, 99% purity.) Abrin came up too, similarly priced. A half dozen such sites pop up when you search for either toxin via its Chemical Abstracts Service number. One vendor advertises free samples for many of its wares — including "horse spleen," though not ricin or abrin.

Pincus hadn't been aware of the website. He used to get his ricin from a researcher at the University of Texas. When ricin was added to the Select Agents and Toxins list, working with it became an extraordinary hassle, and the researcher decided to unload it. She offered her supply — maybe 10 or 20 grams — to the Biodefense and Emerging Infections Research Repository. "They said, 'Great, we'll send someone down to pick it up.' Sure enough," Pincus recalled, "this huge armored truck pulls up with police-type cars escorting it. And now you're telling me I can buy a hundred times that amount on the internet?"

It appears that way, I said. I'll inquire! A day later, two emails from China had arrived in my email inbox. The first one was typed in red letters. It came from someone at the chemical supply finding aid LookChem: "Any information violates the international or national laws . . . is not allowed to be published on LookChem. Once we find this kind of information, we are obliged to report it to the state organs." Its impact was diluted somewhat by the second email in my inbox.

"Nice to contact you," wrote Cathy, a sales manager at the Kaimosi chemical company, to whom the LookChem people must have forwarded my email. "I'd like to avail myself of this opportunity to establish business relations with you." Cathy apologized, saying that they did not have ricin in stock, but offered to make up a custom batch. "Our product is of mature technology, good quality and cheap price," she assured me, "and has been sold overseas and won wide praise." Cathy asked how much I required and when I would need it by.

"I don't know how far you should take this," said Pincus, when I showed him the email. FBI agents used to drop by his office from time to time (where he had a castor plant growing in the corner, unnoticed by his federal callers). Just to chat.

I asked George Smith whether the FBI monitors traffic on these websites. He said they did, but he imagined they'd long ago struck this one from the list. He highlighted the line on the Kaimosi ricin page that

reads, "Store in a cool, dry place." A puri-
fied protein would have to be refrigerated,
Smith said. "If anything, it's minimally
treated castor mash." Most likely I was dealing with the kind of individuals who would
"sell you any white powder for a hundred
bucks a kilogram."

Given the challenges of aerosolizing ricin
and the less-than-genius reputation of the
typical "castor seeds putterer" (quoting
Smith), I imagine the feds don't worry
much about ricin as a weapon of mass
destruction.

According to Pincus, they *should* worry
— though not about ricin in the air or the
food supply. They should worry about this:
"You can take a ricin gene and dump it into
a highly contagious virus, like influenza."
Now you've got a bug that not only infects
millions but also kills millions. (Of course,
you'd first want to have a vaccine to protect
the millions you *don't* wish to kill.) "I've
heard people with homeland defense say,
'Oh, we keep a good control on whatever
genes are synthesized commercially, blah
blah blah.' " Pincus takes no comfort there.
He related another story.

"For therapeutic reasons, we wanted to
synthesize a gene that encoded the toxic
part of ricin, that would be expressed —

i.e., produced — in human cells. This should have set off a lot of red flags if anyone was looking for that kind of thing. But, man, we ordered it, and we had the gene two weeks later. So if you think the Select Agent list protects us from sophisticated terrorists . . ." I'll finish the sentence for him. It's a load of horse spleen.

Also: Terrorists? Try the military of some rogue nation. Try our own nation. Beginning in World War I and carrying on through World War II, the U.S. military experimented extravagantly with ricin. They mixed it in with shrapnel in grenades. They loaded it into four-pound aircraft bombs. They rigged up sprayers with dissolved ricin (or sometimes just castor seed powder). Nothing worked as they'd hoped.

Eventually they shipped some off to the nation's wildlife research labs, one in Colorado and one in Maryland, to try it out on rats.

Warfare and pest control long strolled hand in hand. Both, after all, seek to destroy grouped adversaries as efficiently as possible. Up until the nuclear era, any newfangled deadliness trained upon human enemies also tended to be tried out on enemies of the furred and feathered variety. A United

Nations summary of control efforts used on the African quelea (or "locust bird"), for example, reads like a timeline of military weaponry: "guns, explosives, flame throwers, jellied gasoline, and contact poisons."

During World War II, the chemical warfare people and the agripest control people were united by a common foe: the brown rat. Aka the Norway rat, aka the sewer rat, aka — quoting a Denver Wildlife Research Laboratory press release — "Hitler's ace agent in this country." Supply routes for the raw materials for existing rat poisons had been cut off during the war, and rats were living high: ". . . sabotaging factories, destroying food needed for our allies, and spreading disease among our armed forces." This was not the first time rodents had been portrayed as enemy sympathizers. Posters for a California ground squirrel eradication campaign during World War I featured squirrels in tiny spiked German helmets. "Mrs. Squirrel" has been decorated with the Iron Cross, one of the highest military honors of the German Empire, which she wears as a necklace.

In June 1942, an unusual wartime alliance was forged. Division 9 (chemical weapons) of the National Defense Research Committee (NDRC) of the U.S. Office of Scientific

Research and Development joined forces with the Denver Wildlife Research Laboratory (now the NWRC) in a quest for new rat poisons. The former suggested promising toxins from the arsenal, and the latter tried them out on traitorous vertebrates. Ricin, under the code name "compound W,"* was a candidate, as was sarin. The standout rodenticide, first tested in June 1944, was a phytotoxin the Division 9 people referred to as 1080. It was cheap and quickly lethal to rats.

Long before the ministries of war and agriculture got wind of it, 1080 had been used in rural Africa, in its natural plant form. Then, too, both rodents and humans were potential targets. Because the toxin

---

* Did the makers of the wart-removal product Compound W realize this when they named their product? I don't know, because Prestige Brands, which owns Compound W, doesn't return calls, their online media query form is a dead-end, and they're not on Twitter. But while we're on the topic of inappropriate names, let's consider "Prestige Brands." Because here are some more of their prestige brands: Fleet enemas, Nix for lice, Beano for flatulence, URISTAT, Nōstrilla decongestant, Summer's Eve douche, Boil-Ease, Efferdent denture cleaner, and Boudreaux's Butt Paste.

imparts almost no detectable taste, the aggressor would simply crush the plant and drop it down the enemy's well. While I might question the efficacy of the raw plant material, the lethality of the toxin eventually isolated — a fluoroacetate given the code name TWS — is well documented.

TWS was accidently discovered by a team of Polish chemists who later shared it with Allied intelligence. According to a declassified Division 9 memo dated April 20, 1945, a related fluoroacetate was under consideration as a "contaminant for water supplies" but never used. Personnel quoted in the memo had been shown footage of poisoned dogs and deemed it "an extremely revolting spectacle." (1080 is 17 to 35 times more deadly to dogs than it is to rats.) Both felt strongly that an agent causing death in this "horrible" manner "could not possibly be used by any civilized nation against an enemy even of the most highly depraved type."

So off it went to the Denver Wildlife Research Laboratory to test on rats. One of the laboratory's press releases describes a 1080 trial run at an infested New Orleans grain elevator company. The chemical was dissolved in water and left in half-ounce, rodent tea party–sized cups along known rat runs. Within twenty-four hours, the release stated, 3,690 rats* were dead. Less prodigious but still impressive body counts are given in the 1945 "Summary of Field Reports on 1080," compiled by the NDRC's Rodent Control Subcommittee. One thousand and thirty-two pounds of 1080 was cooked up by chemists at Monsanto and shipped off to rat-besieged army and navy installations and public health departments for field trials.

Along with the kill tallies, the summary details the edibles each location used as bait. These ranged from standard rat bait fare — barley, oats, sweet potatoes, coconut, chocolate, peanut butter — to more culinarily creative concoctions. Naval Base Guam mixed their 1080 with dried eggs, Mazola, and fresh bacon grease, while the Ninth Service Command fashioned a 1080 meatloaf using horse meat and bread

---

* That is a serious fucking rat infestation. *"Huge,"* agreed a spokesperson for GEAPS, the Grain Elevator and Processing Society ("the knowledge resource for the world of grain handling and processing"). The GEAPS man said in an email that even in less developed nations with shoddy storage facilities, "that many rats seems extremely high and uncommon."

crumbs. First Service Command blended 1080 into Meat Hash C-rations. The secret ingredients of the Texas State Board of Health? Popcorn and chicken feed. The summary's authors shared a few of their own favorite bait recipes, complete with cooking directions. ("Blend the 1080 and the flour. Dust the flour-poison mixture over the cubed vegetable, with continuous stirring.")

All was not copacetic, however. Dogs were taking the bait. Or eating the bodies of the dead and dying rodents. And dogs, recall, are exquisitely sensitive to 1080. One test reported fifty had died. The government agencies put their heads together. They made poison lemonade out of lemons. 1080 was registered as a predacide for ranchers to use on coyotes.

The ranchers had a new problem now: how to keep the predacide from also being a man's-best-friend-icide. Especially since afflicted coyotes tended to run off and, in the words of Rodent Control Subcommittee chairman Justus Ward, "regurgitate a sizable quantity of undigested poisoned food . . . over a considerable area." Which ranchers' dogs would find and eat.

Ward turned to Colonel C. P. Rhoads, chemical weapons whiz at the Defense

Department's Edgewood Arsenal. For the predacide version of 1080, Ward politely asked, "Perhaps you could suggest a drug which could be included with the 1080 to reduce the tendency of a coyote to vomit." And for the 1080 rat-killing formulation, might there be something that would do the opposite? An emetic — a vomit-inducer — "so that poison baits which were eaten by dogs would be thrown up much more rapidly than is done when emesis is caused by the 1080 itself." But would that not also make the rats throw up the poison — and thereby survive? It would not, because, to quote a then-secret memo from Birdsey Renshaw, of the National Defense Research Committee of the U.S. Office of Scientific Research and Development, "rats cannot vomit."

World War II ended, but the toxin screening program continued. Over a span of forty-five years, some fifteen thousand potential poisons and repellents were tested by the Denver Wildlife Research Laboratory. Under pressure from environmental and animal welfare advocates, the chemists grew choosier. They looked for poisons that were not just cheap and deadly but also more specific to the animals they sought to control. The poison DRC-1339 seemed to fit the bill, literally and colloquially. The big-flock crop-eaters — blackbirds, starlings, cowbirds, grackles — were all extremely sensitive to it. This was exciting news for the NSA: the National Sunflower Association.

For forty years, the NSA has supported the interests of sunflower farmers, most of whom are in North and South Dakota, smack in the migration path of tens of millions of blackbirds and lesser feasting flocks. You can appreciate the challenge. They are *attempting to keep birds from eating birdseed.* Surveys of blackbird damage to North Dakota sunflower fields between 2008 and 2010 showed average losses of around 2 percent of the crop.

The National Wildlife Research Center has a branch in Fargo, North Dakota, devoted to the sunflower problem. It has been an enduring challenge. They've developed repellents, but there are application problems. Because of the sunflower's unique asana, its "downward-facing head position," aerial crop sprayers end up dosing the undersides of the flowers and missing the seeds. They developed a hybrid with tightly packed seeds that the birds couldn't pull free, but the fat content of the seeds was low. That is a grievous drawback for the sunflower farmer, because the real money in sunflowers comes from oil, not birdseed. (Sunflower seeds make up a modest percentage of most birdfeeder mixes, which is good, because it would be hard to put aside the irony of annihilating birds for a product used exclusively by bird lovers.)

All along the way, the NSA kept up the push for poison. Things came to a head in 2006, when Frito-Lay announced plans to begin frying its major potato chip brands in oil from NuSun, a sunflower variety whose seeds contain no trans fat. To meet the demand, sunflower growers needed hundreds of thousands of acres of new plantings. The NSA sought permission for an increase in the annual "allowable take" of blackbirds. The South Dakota Game, Fish and Parks Department pushed back; ring-necked pheasants, a popular game bird, also turned out to be sensitive to DRC-1339. The National Wildlife Research Center, in 2003, had published detailed data on the sensitivities of dozens of nontarget bird species to DRC-1339, data that the original screening program didn't have or, in some cases, hadn't published. Not just pheasants, but also cardinals, jays, robins, bobwhites, meadowlarks, mockingbirds, tree sparrows, and barn owls showed high sensitivity to DRC-1339. The information has quelled somewhat my enjoyment of the salty, bagged snack foods. *Blood potato chips!*

Killing one or two million of the seventy million blackbirds that descend on the northern plains each year is like trying to solve global warming with an ice machine. The poisoning campaigns seemed almost more like spite than pest control, a practice undertaken out of frustration and anger, rather than documented results. Population modeling undertaken by a team of NWRC researchers in 2002 concluded that if the annual allowable take of migrating black-birds was increased to two million, the benefits to sunflowers farmers "would likely be negligible." Yet the killing continues. In 2018, USDA Wildlife Services destroyed 516,000 red-wing blackbirds, 203,000 grackles, and 408,000 cowbirds.

The irony is that the sunflower growers have long known what works best. As early as the 1970s, contributors to NSA member magazine *The Sunflower* had been recommending effective nonlethal approaches: "primarily habitat and crop management." Give the birds something equally delicious to eat: plant an inexpensive, strategically placed "lure crop." Rather than plowing under the post-harvest plant stubble, leave it for the birds, to distract them from other growers' fields. Apply drying agents so the seeds are ready to harvest sooner, before the flocks arrive. Don't plant sunflowers near marshes with dense stands of cattails or other attractive blackbird habitats. In other words, take the advice of that endlessly quoted general whose name sounds like a new sunflower hybrid:* Know thy enemy.

Lately, habitat management has meant killing cattails. As it had with 1080 back in the day, Monsanto supplied the poison: the contentious herbicide glyphosate (the active ingredient in Roundup). Glyphosate, I just read in a 2012 paper by four NWRC researchers, has been one of the chemicals used to dry out sunflowers for early harvest. Long sigh and downward-facing head.

In war there is always one last option: surrender. "An obvious bird management strategy," wrote George Linz and Page Klug, in their chapter of *Ecology and Management of Blackbirds (Icteridae) in North America,* "is to abandon a bird-susceptible crop and substitute other crops . . . that are not damaged by blackbirds." Or quit farm-

* Sun Tzu.

ing altogether and go into politics, like North Dakota sunflower grower turned senator Terry Wanzek. "We've surrendered," he told Associated Press reporter Blake Nicholson. "The birds won."

9
OKAY, BOOMER:
FUTILE MILITARY ACTIONS
AGAINST BIRDS

As far as I can tell from newspapers of the era, the practice known as crow-bombing reached its zenith near the tiny town of Asa, Texas, on February 6, 1953. Joe Browder, "who hates crows," took 150 pounds of dynamite and parceled it out to make 300 bombs. Half-sticks of explosive were packed, along with metal shards from a local foundry, into cardboard tubes and strung together in the scrub oaks along the Brazos River where the birds returned each night to roost. By one giddy estimate, 50,000 crows lay dead in an instant.

Why didn't someone call the authorities? The authorities were already there. The local game warden was part of the team stringing up the bombs after the crows had flown off for the day to feed. And there was the birds' crime: feeding. A common concern expressed in the newspapers of the time was that crows — "black bandits of the air," "feathered gangsters," the "menacing sable flood" — were raiding the nests of waterfowl, devouring eggs and chicks to the extent that duck hunters would not have enough of them to shoot. Crow-bombing, in fact, was a government-sponsored conservation effort. Who oversaw the bombing deaths of 250,000 Illinois crows during the winter of 1935? The Illinois conservation commissioner. Who offed 3,000 crows at a roost near Coupland, Texas, flew to a wildlife conference in St. Louis, and returned the following Friday to bomb a roost in Creedmoor? The secretary of the Texas wildlife federation.

Such is the inside-out history of conservation in America. It wasn't until the 1980s that the word came to mean what it means to us now. Wildlife and wilderness weren't conserved for their intrinsic value. They were conserved for hunting and fishing. Mighty tracts of wilderness were protected from agricultural and other development to ensure there would always be places and things to hunt and fish. And the ducks were protected from the crows.

Farmers, too, raged against the eating habits of large flocks. In 1952, the agriculture arm of the government gave dynamite a try. The Denver Wildlife Research Laboratory ran a series of "experimental bombing tests" in the trees along a mile-long Arkansas marsh where redwing blackbirds, grackles, starlings, and cowbirds would come to roost after a long day of rice pilferage.* The study aimed to compare the cost-effectiveness of different bomb materials: dynamite versus Primacord, lead shot versus steel, pasteboard mailing tube versus ice-cream carton versus can. Because of this work, we now know, for instance, that the "average kill per bomb" for dynamite and lead shot packed into a mailing tube is 1,820 birds, at a "cost per bird killed" of less than a cent.

What the experiment did not reveal was

* Let us here lay to rest the myth that birds' innards will burst if they scavenge the rice thrown at weddings. Over the years, this enduring bit of misinformation migrated as far as Ann Landers's column and the Connecticut state legislature. In 1985, Representative Mae Schmidle proposed "An Act Prohibiting the Use of Uncooked Rice at Nuptial Affairs." The Audubon Society called hooey, pointing out that migrating birds feed on fields of rice. Some churches ban the practice anyway, not because it's perilous for birds but because it's perilous for guests, who could slip on the hard, round grains and fall and then fly off to a personal injury lawyer.

the extent to which killing what amounted to 1 or 2 percent of an enormous flock had a noticeable effect on farmers' losses. That assessment in fact already existed, in the form of data collected over a decade of bombing crows in Oklahoma. From 1934 to 1945, 127 crow roosts were dynamited "to reduce predation on waterfowl eggs and damage to grain crops," writes longtime USDA ornithologist Richard Dolbeer in his chapter of *Ecology and Management of Blackbirds (Icteridae) in North America*. An estimated 3.8 million crows were killed, yet, Dolbeer writes, "no evidence was obtained to indicate the explosions influenced total population levels, agricultural damage, or waterfowl production."

Here's the thing with killing as a wildlife damage control tool. It isn't just mean. It doesn't — barring wholesale eradication — work. On a visit to the National Wildlife Research Center, I spent a morning in the archives, reading oral history transcripts of some of the long-timers. One in particular stayed with me. Weldon Robinson began his career as a government bounty hunter, killing coyotes for $3 a head. He was soon hired by the Denver Wildlife Research Laboratory, where he rose through the ranks. By 1963, Robinson oversaw four units, including Bird Control, Predator Control, and Agricultural Rodents. He was the czar of nuisance wildlife offing. Somewhere along the way, Robinson's oral history reveals, he had an epiphany regarding the animals whose numbers the agency had for so long fought to reduce: "Mother Nature adjusts."

Robinson had landed on the phenomenon of compensatory reproduction. Destroy a chunk of a population, and now there's more food for the ones who remain. Through a variety of physiological responses — shorter gestation periods, larger broods, delayed implantation — a well-fed individual produces more offspring than one that's struggling or just getting by. With ample food, both the well-fed parents and their well-stuffed young are more likely to survive and reproduce. Coyotes, for instance, may have three pups when food is scarce, and eight when it's plentiful. That figure comes from *Resolving Human-Wildlife Conflicts,* a technical press book by Michael Conover. Conover is a former director of the Jack H. Berryman Institute, which sponsors research into methods of human-wildlife conflict resolution (plenty of them lethal). Conover adds that killing coyotes opens up territories for males that would otherwise not breed. Bottom line: in order to end up with a net reduction in a coyote population, humans would have to destroy at least 60 percent of them annually.

Robinson shared his catchphrase for it. "Natality is more effective than mortality," he said to the interviewer. Catchy, but a bit abstruse. She changed the subject. He went right back to it. Here I pictured him leaning forward in his chair. "Natality is more effective than mortality," he repeated. Whereupon the interviewer said she had no more questions, and they looked at some old photos and marveled at how a colleague named Johnson Neff used to have hair.

Another thing I ran across in the oral histories was a mention of the practice of overestimating flock sizes in order to secure federal funds for bird control. The center's head of bird depredation control was explaining to the interviewer how he'd estimate the size of a flock as it flew between two points overhead. The technique wasn't perfect, but it was, he said, better than just making up a big number, as some states were doing. " 'There's twenty million birds! We need more money.' [*Chuckles.*] Yeah, the farmers would do the same thing. 'I got two hundred thousand birds in my field!' I need more! [*Chuckles.*]"

It wasn't even clear, in the case of black-birds eating rice, that the take represented a significant loss. By the estimates of a 1971 Fish and Wildlife Service Resource publication, the average loss per acre was between half a bushel of rice — by volume, about four gallons — and just over one bushel. That is less, the document goes on to say, than the amount that would typically fall to the ground as a combine harvester moved through the crop (a loss of 1.6 to 8 bushels per acre).

And then there's this: birds provide significant pest and weed control services to the farmer. Based on an examination of nearly five thousand blackbird stomachs, Foster Ellenborough Lascelles Beal of the Bureau of Biological Survey (a forerunner of the NWRC) had this to say: "Judged by the contents of its stomach alone, the redwing is most decidedly a useful bird. The service rendered by the destruction of noxious insects and weed seeds far outweighs the damage due to its consumption of grain." The report was signed by none other than the secretary of agriculture. Today you have to go to organic farming organizations* to

* Or Mormon history books. In 1848, a flock of California gulls flew over from the Great Salt Lake

253

find this sort of information. (The Wild Farm Alliance, for instance: Based on stomach contents, "each Tufted Titmouse is worth about $2,900 to the pecan industry.")

Bird-bombing was soon abandoned, Dolbeer wrote in the blackbird book, "for various obvious reasons: labor, expense, hazards involved, large rate of crippling . . . and lack of effectiveness in solving problems." It was odd, then, that the Arkansas bombing experiment report concluded with a thumbs-up. The results, stated researchers Johnson Neff (Neff! With hair!) and Mortimer Brooke Meanley Jr., indicated "the effectiveness and economy of roost bombing."

Were Meanley and Neff, like Joe Browder, simply men with a hatred of crop-raiding birds? And/or a love of setting off bombs? Bird-bombing seemed to peak after World War II. I began to wonder if it was the product of some lingering zest for combat, some wartime hangover of misplaced patriotic zeal. An obituary for Meanley put that notion to rest. His World War II service took the form of rehabilitating soldiers by taking

---

and descended like the hand of God upon swarms of insects devouring the settlers' crops. Which explains why to this day, the state bird of Utah is the California gull.

them out into nature. "He would comment on the unbelievable good fortune of meeting his military commitment by taking soldiers on bird walks." Both Meanley and Neff were respected ornithologists.

How, then, did this gentle bird lover, author of *Natural History of the Swainson's Warbler,* end up dynamiting blackbirds in an Arkansas bayou? The same way, I suppose, that entomologists end up developing insecticides and wildlife biologists find themselves having to put down bears. The jobs are few, and that's where they lie. For someone whose know-how is birds, their control was one of few ways to make a living.

I can't say why Meanley and Neff endorsed bombing. Perhaps they never read Mr. Beal's blackbird report or the paper on the lackluster outcome of the Oklahoma crow-bombing campaigns. It would seem no one read Neff and Meanley's paper, either, because blackbird control quickly shifted away from bombs.

And over to chemical warfare. Five years later, Neff and Meanley were out in the fields again, scattering strychnine-treated grains around the roosts of blackbirds and cowbirds. Both species, wrote Dolbeer, "generally avoided the baits." Or maybe just avoided Johnson Neff and Mortimer Brooke Meanley.

The war on thieving birds has occasionally strayed beyond the realm of metaphor and taken the form of an actual military operation. In October 1932, the Australian minister for defense agreed to dispatch a pair of machine-gunners, led by a Major G. P. W. Meredith, to rout mobs of emus trampling and helping themselves to farmers' wheat in the state of Western Australia. (The defense minister had denied the farmers' original request, which was to borrow some machine guns.) In return, the military asked only for the feathers of the fallen, to be used as a flourish on the hats of the nation's light-horsemen.

The emu proved a tougher adversary than Major Meredith and his gunners had anticipated. Though flightless, the bird is a fast sprinter, reaching 30 mph with proper motivation. Emus blend with their surroundings and travel with a lookout that, in this case, would sound a warning long before the birds were within range of the guns, whereupon the flock would scatter in a rising billow of dust. On day three, with a confirmed body count of just twenty-six birds, Major Meredith shifted tactics. He set up an ambush, concealing his gunners in the scrub above a dam where emus would come to drink. Around 4:00 p.m., a sizable flock was spotted in the distance.

"The constant craning of necks and the cautious manner of their approach showed they had not forgotten the happenings of the last few days, but relentless thirst drove them on," wrote a correspondent for the *West Australian*. When the birds were a few hundred yards out, Major Meredith gave the order to fire. As the dust settled, the men got up to count the bodies. An underwhelming fifty birds lay dead. Excuses were made. The machine gun jammed, someone told a reporter. Someone else conjectured that the majority of the bullets were passing harmlessly through the birds' plumage, because the emu has "more feathers than flesh." Major Meredith believed hundreds more had been hit but survived. He credited the emu with an almost supernatural ability to "face machine guns with the invulnerability of tanks." He sounded wistful. "If we had a military division with the bullet-carrying capacity of these birds it would face any army in the world."

On day six, Major Meredith withdrew in defeat. "Emus appeared in huge flocks along the road," observed the Perth *Daily News*, "as if to give a mocking farewell." And that was that, almost. Twelve years later, as World War II raged, the upstart wheat farmers of Western Australia would again ask for military intervention. This time they wanted "light bombs dropped from low-flying planes." The request was denied.

Meanwhile, out on the Pacific Ocean, the albatrosses of Midway Atoll were shaping up to be an equally insuperable foe.

The islands of Midway sit halfway between North America and Asia, in the Pacific Ocean. As such, Midway Atoll was strategically important to the United States, and in 1941, the nation built a naval air station there. Midway was (and is) also important to a dozen or so species of seabird, including the tens of thousands of Laysan and black-footed albatrosses that return to the islands each year to lay eggs and raise their chicks. Because the birds had had no predators on the islands, they greeted newcomers — both human and mechanical — not with fear but with a mix of nonchalance and curiosity. They soared over the navy's runways with blithe disregard for the large, loud metal birds with which they shared airspace. Collisions — bird strikes — became an issue.

"We got a bird in the carburetor air scoop." An aviator machinist named Jerry is speaking into the microphone of a navy public affairs man in a 1959 government filmstrip. "Caused a total loss of power in our number three engine."

The microphone tilts back to the interviewer's mouth. His moustache is trimmed to an inverted V, a soaring upside-down albatross of a mustache. "What would happen if you took a gooney bird in the air intake of a Super Constellation radar plane?" ("Gooney bird" was the de facto military handle for albatrosses.) "Do you think you would crash on takeoff? Total loss of plane and personnel?"

"Yes, sir, more than likely."

The interviewer turns to face the camera. "And there you have it. From people who know what they're talking about. It's hard for them to understand the continued existence of these gooney birds here on Midway."

The scene shifts to an office at Naval Air Station Barbers Point on Oahu. We are introduced to Rear Admiral Benjamin E. Moore and his pointer. Admiral Moore stands beside an easel on which is propped a posterboard highlighting key gooney bird cost statistics. "We had five hundred thirty-eight strikes last year." The pointer taps the text BIRD STRIKES 538. He then breaks down the cost of these strikes, beginning with the wages of the men who repair the damage. "Two thousand five hundred twenty man-hours times two dollars an hour equals five thousand forty dollars." Admiral Moore steps to a second easel. The pointer taps FLIGHTS ABORTED 33. Each aborted flight requires the pilot to dump three thousand gallons of fuel to reach a safe landing weight. The pointer directs our gazes to JETTISONED FUEL 99,200 GALLONS and below that, $17,500 COST. The admiral returns to the first easel, where someone off-camera has snuck in and changed the posterboard. A final, decisive tap of the pointer: $156,000 TOTAL.

Admiral Moore steps to his desk. To his left, an American flag hangs on a pole in the sad, limp manner of all indoor flags. The admiral sits. The mood darkens. "It's either remove the gooney birds from Midway and keep the men, or keep the birds and eventually bury a flight crew of twenty-two men. I hope I'll never have the task of explaining to some mother or wife that her son or husband was killed by . . ." Admiral Moore pauses to raise a photograph that the easel sprite has slipped onto his desk. He glowers over the top of an enlarged photograph of an albatross standing calmly on a lawn. *". . . this."* A dramatic orchestral swell accompanies a fadeout and the words THE END.

The filmstrip is titled *The Second Battle of Midway.* And the battle was a lengthy one, longer than the first, longer than World War II itself. In the beginning, the strategy was straightforward massacre. Because the birds were many, the cost of ammunition high, and the admiral budget-minded, the initial assault was done without guns. It took place in 1941 and is detailed in a Fish and Wildlife Service Special Scientific Report, under the heading "Large Scale Elimination Experiment." Two hundred men carrying lengths of pipe or wooden clubs "spent 6 or 7 hours a day" hitting albatrosses on the back of the head. An estimated eighty thousand birds were killed. "For a brief time the hazard to aircraft was reduced," the paragraph concludes. "The following season there appeared to be as many albatrosses as before."

The strategy shifted to harassment. To persuade the albatross to nest elsewhere, personnel fired off everything they had. Under the heading "Disturbance," a 1963 government review of albatross control efforts at Midway lists "rifles, pistols, bazookas, mortars." "Some birds registered discomfort" but "no appreciable number" left their nests. Ten carbide exploders were set up at intervals of 135 feet along the edge of the runways where the nests were densest. "No decrease in number of flying birds." Personnel burned rubber tires and lit flares to roust the birds with noxious smoke. They tried blasting an "ultrasonic siren," thinking, mistakenly, that the birds could hear ultrasound. At one point a Lockheed WV-2 Warning Star radar surveillance aircraft taxied to within a hundred feet of the nest area and switched on a high-intensity radar beam. With no observed effect.

Unable to harass the birds off the island, the navy entertained the possibility of physically moving them. In another test program, eighteen albatrosses were plucked from the nesting site near the runways, banded, and put on outgoing military transport flights: to Japan, the Philippines, Guam, Kwajalein, and — *aieeee!* — Naval Air Station Barbers Point on Oahu, nesting site of the military's highest-ranking albatross hater, Admiral Benjamin E. Moore. Fourteen of the eighteen flew (sans airplane) back to Midway in time for nesting season. The navy had not realized that although an albatross wanders thousands of miles, it always returns to the same spot to nest.

And no, you cannot just move the nest. The navy tried this, too. The birds would consult their internal GPS, see that the nest had been moved, go back to their actual birthplace, and proceed to construct a new nest. Next the navy tried moving the eggs. Ten thousand albatross eggs were carried to the adjacent island, where the sailors chased the local birds off their nests and quickly installed the kidnapped eggs. It was like those Folgers ads from the 1970s. *We've secretly replaced these birds' eggs with eggs from the neighboring island.* The birds, unlike the instant-coffee dupes, were not fooled.

Exasperated, the navy turned to science for help. On October 2, 1957, Hubert Frings, a Pennsylvania State University professor of zoology, got a call from Washington. Would he be willing to come to Midway Atoll for the December/January nesting season? In other words, would Frings make the supreme sacrifice of spending his between-semester break on a tropical island in lieu of hanging around Altoona, Pennsylvania? You bet. Accompanying him would be his wife, Mable, a librarian and bioacoustician with a special interest in cricket and grasshopper "chirp sequences."*

On the Fringes' very first day at Midway, another "large-scale elimination" was taking place. Despite the documented failure of earlier mass killings, there was talk of clubbing the entire Midway population. Mable went out into the fray with her reel-to-reel, thinking to record the albatrosses' alarm and distress cries, so that these might later be played in a more ornithologically informed effort to scare them away. There were no cries to record. "The birds," Hubert wrote in his memoir, "mostly sat quietly in place until clubbed." Mable's tapes captured only "the crack of ash on skulls" and, at one point, the distress cry of a young sailor: "I didn't join the Navy to crush the heads of innocent birds."

Hubert quietly raised the issue of morale and, if word got out, "the reaction of people

* It was Hubert and Mable Frings who debunked the folksy almanac claim that you can determine the temperature outside — or, if you are the Frings, inside your own apartment — by counting cricket chirps. (Thusly: Take the number of chirps in 25 seconds, divide that by 3, add 4, try to recall if that's the Fahrenheit or the Celsius formula, and add "download weather app" to your to-do list.)

stateside." His concerns were ignored. As before, the slaughter failed to meet its objective. Though 21,000 birds had been killed, "there were," observed Hubert, "almost as many still wandering around near the runways, and counts of bird strikes by planes remained the same." Besides, he continued, "even if total annihilation on Midway were achieved, it would probably be only a short time until the land available sign would bring in new settlers."

Hubert and Mable did what they could to provide kinder alternatives. They suggested clearing sea grape from the beaches on nearby islands to make them more appealing to house-hunting albatrosses. The plan has a cameo in the navy publicity film. We see Admiral Moore's pointer tapping a photographic blowup of Kure Atoll. The Navy Bureau of Aeronautics, he states in the film, planned to raze the shrubbery, but I could learn nothing more about it.

The navy did try its hand at "terrain modification," though this was on Midway itself. Someone had suggested that a row of dunes near the runways was creating updrafts and that the albatrosses needed these to fly. And that leveling the dunes would therefore solve the problem. Hubert didn't think so. The wind off the ocean, combined with the sizable surface area of albatrosses' wings, provided all the lift required. He pointed this out, but the dunes were bulldozed anyway. If anything, there were now *more* albatrosses soaring over the runway, as the demolition opened up easier access to the area.

The weeks passed, and Hubert had to return to his teaching. He and Mable agreed to do some experimenting back in Pennsylvania. The pair were keen to develop an albatross repellent. The navy, by now experienced shippers of live albatrosses, promptly dispatched two to the Fringses. A photograph in Hubert's memoir shows Mable in a short-sleeved dress and two-tone flats, lifting the lid on a plywood crate. Poking up over the rim are the heads of two Laysan albatrosses fresh from the final leg of their journey, the express train to Altoona. The birds appear, as ever, utterly unfazed by the strange things humans do to them.

The albatross-repellent trials were a disappointment. Mothballs did not bother them. Live snakes were presented and ignored. Recordings of distress and alarm calls likewise caused no reaction. This was more or less expected, because in order to elicit and record these calls, a Midway albatross was swung in a circle, and while this went on, nesting birds a few feet away "did not even look over to see what was causing the commotion." The albatross is a largely unflappable bird.

The Fringes returned to Midway the following January. They were running out of ideas. One day Hubert noticed a group of navy wives effortlessly shooing a flock of loafing albatrosses by walking onto their lawns holding tablecloths in front of themselves. "We made tests with various surfaces held in front of us, walking toward the birds," he wrote. ". . . A sizable flat surface was . . . quite repellent." Excited by his findings, Hubert set a meeting with Midway command. With an "integrated program" of men holding large colored squares, he believed, they "might very well be able to clear the island of nesting birds." He estimated that twenty to thirty personnel would be needed, daily, for the duration of the albatrosses' return to their nests.

The proposal was not embraced. It had followed on the heels of a proposal to harass the albatrosses by stringing low wires across the nesting area — by tripping them. Hubert's final suggestion was inspired by an observation that no albatross was ever seen flying under the metal aprons suspended from some of the hangar roofs. He wondered whether suspending long panels of fabric might stymie any albatross seeking to nest where it oughtn't. He envisioned colored panels 10 feet long by 3 feet wide, suspended between 20-foot poles in the beach along the runway. It sounded marvelous, if you were a wedding planner, or Christo. Less so to a navy pilot. Hubert broached the idea at the Officers' Club one evening. "Opinion," he wrote in his memoir, "was not unanimous on this."

It was around this time that attitudes toward the Fringses began to deteriorate. "Our work," he recalled, "was regarded . . . as useless and troublesome." And so Hubert Frings returned to his teaching, and Mable to her grasshoppers and crickets and to a new project, "reviewing the life styles of spiders." I end with a quote from Hubert's journal: "Of all the animals we have ever kept, these Albatross are the ones to which I have become most attached. I really love them and respect their independence and jaunty ways. This marks the end of a period of acquaintance with real aristocrats of life." I kind of feel that way about Hubert and Mable Frings.

The navy closed Naval Air Station Midway

in 1993. No plane ever crashed. No airmen died because of the gooney bird. Bird strikes had continued throughout the navy's tenure, despite everything it tried. There were, one report stated, twice as many collisions at the end of one four-year albatross control program as there had been at the time the effort started.

In September 1958, *Flying* magazine ran a piece about Midway's albatross conundrum. An airman was quoted saying, "It's my bet that no matter what we throw at them, when it's all over, Midway will still be the Gooney territory and ourselves only transients subject to the whims of a bird that refuses to be conquered."

I hope the airman put cash on that bet. Because it is all over now, and the islands are very much gooney territory. Naval Air Station Midway is now Midway Atoll National Wildlife Refuge. Nothing goes on there but the happy hatching and rearing of seabirds and the quietly atoning efforts of Fish and Wildlife personnel, now working to restore the birds' habitat.

All around the world, wildlife continues cluelessly, tragically, to cross paths with large vehicles. And science continues, sometimes entertainingly, always earnestly, to search for solutions.

10

ON THE ROAD AGAIN:
JAYWALKING WITH THE ANIMALS

On July 26, 2005, the space shuttle *Discovery* hit a turkey vulture. The encounter, which happened during liftoff, was caught on video. You see the great bird soaring, perhaps delighting in the awesome thermals of a rocket liftoff, and then abruptly bouncing off the external fuel tank and plummeting, Icarus-like, into the exhaust plume. Because the craft had just begun to accelerate, the damage was small, and largely confined to the vulture. Nonetheless, recalls Travis DeVault, a wildlife biologist with the Sandusky, Ohio, branch of the National Wildlife Research Center, "it created a lot of heartburn."

The NWRC branch offices share the grounds of NASA's Plum Brook Station. Plum Brook is where NASA tests rocket engines and Mars landers and the like to be sure they'll perform under the stresses and unusual circumstances of space travel. Plum

Brook engineers can create wind that blows six times faster than the speed of sound, and vibrations as ungentle as those of a rocket launch. A wayward buzzard seems laughable by comparison, but no one was laughing. Travis reminds me that it was a ding to space shuttle *Columbia*'s exterior on the way up that led to the tragic explosion on the way back down.

Plum Brook Station is headquarters to the National Wildlife Strike Database, co-managed by the FAA (Federal Aviation Administration) and the USDA. Plum Brook is federal chop suey. The FBI is here too, behind a row of closed doors with nameless nameplates, across from Travis's office. He doesn't know what they do in there but is impressed by their paper-shredding capabilities. "Comes out like *dust.*"

In 2015 the National Wildlife Strike Database summarized data from twenty-five years of collisions between civil aircraft and wildlife. The report goes species by species:* number of strikes, the total cost of

---

* How is it possible to identify the species of a bird that has slammed into a nose cone or passed through the whirling blades of a jet engine? Forensic ornithology! Feathers, down, beaks,

damage to aircraft, and the number of people killed and injured. There are two ways to be a bird of interest to the FAA (or, who knows, the FBI): weigh a lot or travel in groups.

By dint of their size, turkey vultures* rank high on the worry list: eighteen people injured, one killed, and costly damage to the aircraft 51 percent of the time. By contrast, of twenty-seven documented

---

talons, and/or "snarge" (tissue scraped from the plane) are collected and shipped to the Smithsonian Institution's Feather Identification Lab. The United States Postal Service happily delivers snarge. They also, I learned on their website, will deliver live animals that are intact — including leeches, goldfish, scorpions (double container required), birds lighter than twenty-five pounds, and "small harmless cold-blooded animals," whom they have also hired, at least once, to man the counter at my local post office.

* To clarify, a turkey vulture is a vulture, not a turkey. Though turkeys, too, have crashed into planes. But only wild ones. Supermarket turkeys have never hit planes, but supermarket chickens have, because they are fired at jet parts to test their ability to hold up to birds strikes. The device that fires them is called, yes it is, the chicken gun.

chickadee strikes, none caused damage of note.

Jet aircraft engines are given a "bird ingestion" test, but the birds used in the testing weigh two pounds. Turkey vultures average three. Travis sent me a link to footage of one of these tests, slowed down to reveal a progression of fan-jet engine blades slicing a bird like a meatloaf. With a bird the size of a vulture or a pelican, the blades may end up in pieces, too. Shards of fan blade slamming into delicately calibrated engine parts can have catastrophic consequences.

Travis also emailed me control tower footage of a Boeing 757 strike that he uses at trainings for airport biologists. You see an unidentified blackish bird — really just a speck on the screen — disappear into the mouth of an engine during takeoff and produce a near-instantaneous fart of fire out the back. It's the soundtrack that stays with you. You hear the pilot's call, "Mayday, Mayday, Mayday," and in the background the normally cheerful but now darkly sinister twitterings of a bird.

Odds are decent it was a starling — among the six most frequently struck bird species in the United States. To confuse winged predators, starlings sometimes fly in enormous shape-shifting flocks called mur-

murations that swerve and split and coalesce again, all without warning or seeming logic. The open maw of a jet engine passing through is bound to swallow a few. They're like avian krill.

Worst-case scenario: big birds in groups. It's no surprise that the species that brought Captain Chesley B. "Sully" Sullenberger down to the Hudson River was Canada geese, one or two per engine.

Travis DeVault's wildlife strike research has focused, at various times, on turkey vultures, blackbirds, and Canada geese, but tonight he'll be gathering data on a yet more dangerous creature, the species the FAA considers the "most hazardous wildlife to U.S. civil aircraft": the white-tailed deer.

From 1990 to 2009, the National Wildlife Strike Database logged 879 collisions between white-tailed deer and aircraft. The impacts injured 26 people and caused, on average, six times as much aircraft damage as other wildlife strikes. Only Canada geese, red-tailed hawks, and pelicans caused more aviation deaths. Deer strikes have happened on landing, during takeoff, and while taxiing. Obviously no deer have struck planes at cruising altitude. Less obviously, two animals struck planes that were parked.

White-tailed deer are thirty times as heavy

as turkey vultures, and they travel in groups. Double whammy, both for aircraft and for vehicles on the road. Travis's focus, of late, has been this: Creatures on roads and tarmacs often don't get out of the way in time, even when there's time to get out of the way. He's looking for answers to the questions we can't just ask. Why do you stand there in the headlights? How can we help you? How do you *not* notice a space shuttle?

Travis is driving me around Plum Brook's six thousand woodsy acres. It's a wildlife biologist's tour of a space center: *See the bald eagle nest up there? And this is a great place to hunt for mushrooms. That building has some big thermal thing. There's another nest!* You can tell which buildings are rarely used, from the deer. A half dozen of them mill about on the lawn of the Hypersonic Tunnel Facility, nibbling and strolling like guests at a wedding reception. Clusters of them appear along the roads with the regularity of mile markers. For the first couple minutes, I'd lean forward against my seat belt and exclaim, "There's one!" At one point Travis turned to me: "You don't see many deer where you live, do you, Mary?"

Out where he lives, Travis sees a lot of

deer, but he did not see the one he hit. This is often the case. "The deer you hit are the followers," he says. You brake, and your eye follows the deer you just avoided hitting. "You speed up, and here comes the next one."

As soon as it's dark, we'll be meeting up with Travis's research partner, Tom Seamans, to take some data for what I'm calling the deer in the headlights project.

It's approaching dusk now, the time of day with the highest hourly incidence of deer-vehicle collisions — three times higher than the dark of night, according to a paper by Travis and four colleagues. Deer are crepuscular, a word born for dermatology but in fact meaning "active at dawn or dusk." November is the other standout risky drive time, because it's the rut — mating season. Hell-bent on reproduction, the deer fail to note the most blatant obstacle to the successful onward advancement of their genes: traffic.

We're seeing deer all along the roads partly because it's dusk, and partly because Plum Brook has a lot of them: about ten deer for every one person who works here. Also, a road that runs through woods is an attractive place to a deer. Food grows close by, and it's a clearing of sorts, so predators can't easily sneak up unseen. The open space of a road also appeals to birds that hunt insects on the wing, because it's easier to see and maneuver here. The ones that get hit draw the scavengers. Roadkill begets roadkill. (In hopes of preventing another turkey vulture mishap, the Kennedy Space Center grounds crews formed a "roadkill posse" to clean up carcasses with unusual alacrity in the days leading up to a launch.) On the simplest level, animals take to the road for the same reason people do: the going is easier.

The low speed limit at Plum Brook has helped keep wildlife populations booming. As long as cars aren't moving faster than a natural predator would, a prey animal will usually get out of the way in time, even if the driver doesn't brake. The hunted maintain what's called a spatial margin of safety. They're able to visually intuit the distance between themselves and a predator, and they have an uncanny sense of exactly how close they can let that predator come before they need to take off. Flight initiation distance, or FID, as that closest point is called, shrinks and lengthens according to circumstances. If animals or birds are feeding on something nutrient-rich and wonderful, they may wait until the last possible moment — the shortest FID — to abandon the feast. If the predator is coming at them at a run, its speed is factored in and they'll take flight when their pursuer is farther away. They almost always judge the safe getaway distance correctly. Unless the thing coming at them has an engine.

Mammals and birds, sensibly, perceive onrushing cars as predators. Their escape algorithms work well on a congested city street — though you may try, you will almost never hit a pigeon — but their judgment fails them on a freeway or a rural thoroughfare. Because what predator comes at you at 60 mph? Evolutionarily speaking, this is something new. "Fast cars have only been around for a hundred years," Travis says. He flips a visor down against the setting sun. "In terms of evolution, that's nothing."

Travis has speculated that this explains the baffling inability, among wildlife, to avoid what should be easy to avoid: "a large, noisy vehicle traveling along a predictable path." Evolution hasn't had time to upgrade the processors. Judging speed requires an ability to perceive and interpret "looming" — how quickly an object appears to be growing in size as it comes at you. Looming is harder to detect and visually process when the object is traveling quickly. The "looming-sensitive neurons," as some pigeon researcher went ahead and named them, are overwhelmed.

Travis and Tom have devoted considerable time to studying this. The pair's original study protocol was straightforward: "We drove a vehicle directly towards turkey vultures . . ." The vultures had been lured by a raccoon carcass anchored to a heavy metal plate, to keep the birds from dragging it off the road to a more relaxing setting. The vehicle, a Ford F-250 pickup, traveled without braking, at three different, constant speeds: 19, 37, and 56 mph. Flight initiation distance was measured by dropping a beanbag out the window onto the asphalt to mark the moment a vulture moved to retreat, and then measuring the distance from beanbag to carcass. The FID for 37 mph was not significantly different from the FID at 56 mph, suggesting that, as Travis and Tom had predicted, an unnaturally fast "predator" overtaxes the prey's sensory and cognitive gifts.

No vultures were hit, though there were close calls, all of them at the fastest speed. To see what would happen at even faster speeds, Travis and Tom devised a video truck. Cowbirds — because they're common around here and hardy — were installed in a roomy cage (and, fret not, released afterward). One wall of the enclosure took the form of a video screen, on which the researchers played footage of a truck driving straight at a video camera they'd placed in the middle of a road. Cowbirds, they found, will take to the air when a vehicle gets to about 100 feet away, regardless of its speed. Up to about 75 mph, they have ample time to get out of the way.

Through the magic of variable speed video playback, Tom and Travis were able to accelerate the video truck to 224 mph — roughly the speed of a plane taking off. Because that's really what this research is about: flight safety and airplane damage prevention. It would be great to figure out how to prevent the deaths of hundreds of millions* of small creatures on U.S. roads each year, but that's not the ultimate goal

---

* Or probably more. Figures for roadkill prevalence tend to be underestimated owing to the speed with which scavengers arrive to eat the evidence. Within twenty-one hours, one Mojave Desert survey found, all that remained of a crushed tortoise on the centerline of the road were scattered pieces and "two withered limbs eight meters from the point of impact." At ninety-two

of the work.

Faced with a truck going as fast as a plane, every one of the cowbirds would have wound up a statistic in the National Wildlife Strike Database.

Research on jaywalking pedestrians tells a similar story. Most of our decision-making is based on how far off a car is. We're not so good at factoring in the speed. Experimental evidence suggests that full looming sensitivity doesn't develop until adulthood. A young child on the side of the road and a car traveling faster than 20 mph combine to encourage, quoting a team of European psychologists, "injudicious road crossing." Hence the need for injudiciously punctuated SLOW CHILDREN signs. It's not just that kids aren't looking when they cross; they're also not seeing.

For animals facing down a predator, fleeing is but one option. Mammals rely on a diversity of features and behaviors that, over the millennia, have increased their odds of staying alive long enough to pass on their genes. The skunk sprays a vile smell, the porcupine wears darts. When the "predator" is a speeding automobile, these tactics

---

hours post-impact, the tortoise was but a "faded stain."

range from ineffective to tragically maladaptive. The turtle stops in its (and your) tracks and pulls its head into its shell. A deer may freeze to avoid being seen among the trees. Squirrels and rabbits zigzag and spazz halfway across a street. When your killer is a hawk that has calculated the likely intersection of its path and yours, changing course abruptly may save your life. When the killer is a land-based commuter, it foils her efforts to avoid hitting you.

When self-driving cars take over the roads, the lives of squirrels and skunks (and cats and small dogs) may no longer be spared by kindly drivers who swerve and brake. By the cold calculus of (human) survival, drivers are safer doing neither. The Centers for Disease Control and Prevention estimates that ten thousand people per year are injured when they take evasive action to avoid hitting an animal. That's only two thousand fewer than the number injured when the vehicle actually hits the animal. In 2005, the Insurance Institute for Highway Safety (IIHS) analyzed 147 fatal (to the human) vehicle-animal crashes, 77 percent of which involved deer. The initial impact rarely killed or even hurt people; almost always, they died because the driver tried to avoid the animal. The driver or motorcyclist braked, and the vehicle skidded and went off the road or collided with something less yielding than venison.

The eight exceptions were instances in which large deer — and in one case, a horse — crashed through the windshield. Taller is a killer. Because now the car's front end strikes the animal's legs rather than its torso. And when the legs are knocked out from under, the torso and head pinwheel over the hood and crash down onto the windshield and, if the animal is tall enough, the roof. Thus Volvo has a LADS — large animal detection system — but no SADS. "The camera looks for a specific signature," a Volvo communications manager said in an email. "A large body mass with four very thin long legs." The example he gave is a moose.

For a 1986 master's thesis, a team of Swedish bioengineering students staged a moose strike and filmed it at high speed, so the impact and its aftermath could be studied in slow motion. The aim was to provide a more nuanced understanding of the biomechanics of these often devastating collisions, and then use this understanding to develop a moose crash test dummy. An "ill and weak" bull moose "was put to death and quickly afterwards it was hit by a Volvo

240 at a speed of [50 mph]." The phrasing intrigued me. Apparently the Volvo 240 is a car that goes from 0 to 50 quickly enough to reach a moose in the fleeting moments between its death and the crumpling of its legs. For you could not suspend a carcass in a frame, as this would keep the moose from doing the very thing the authors aimed to study.

Anyway. What the film revealed. If the roof collapses as the passenger is thrown forward — and now I borrow the gently evocative phrasing of Swedish moose crash test dummy designer Magnus Gens — "crumbled steel interferes with the head's path." Borrowing the less gentle phrasing of "Moose and Other Large Animal Wildlife Vehicle Collisions," "axial compression . . . causes bony fragments to be pushed into the spinal canal." Moose falling on driver's head crushes neck vertebra, sharp pieces of which slice spinal nerves, causing full or partial paralysis. Also disturbingly common: broken face bones and lacerations from hitting the windshield in mid-shatter. The wounds become infected by "debris, hair, entrails, and feces." And finally, should the two of you manage to survive the impact, one of you now has a flailing moose in his lap.

Making matters worse: the long legs of a moose may boost the animal's eyes up above headlight range, eliminating the reflective shine that helps drivers see it in the dark. (The tapetum lucidum, which does the reflecting, is actually there to aid *their* vision, not ours. It boosts mammals' vision in low-light conditions by bouncing the light back into the retina a second time.)

If you plan to be driving in far northern regions where tall ungulates are likely to dart into the road, you might want to consider a Saab or a Volvo, as their roof pillars and windshields are designed and reinforced with input from Magnus Gens's awesome moose crash test dummy. Magnus received funding from the Swedish National Road and Transport Research Institute, which shortly thereafter received a plea to help design a camel crash test dummy.

A camel is taller and heavier, and therefore even deadlier, than a moose. More of the roof is likely to collapse directly onto drivers' heads. If they lean or duck sideways to dodge the hurtling ungulate, now their back is likely to be broken instead of their neck. Of sixteen Saudis whose cars struck camels, one study relates, nine wound up with complete quadriplegia. Along certain stretches of highway, camel density has been as high as nineteen per mile. These animals are not wild, but their owners often allow them to roam. Sometimes even, in days past, encouraged it. Because until recently, Saudi law required the driver to pay the camel's owner for the loss. "Therefore," reports a team of neuroscientists at Riyadh Armed Forces Hospital, "some camel owners have been known to push their animals onto the highways after sunset to claim compensation after the accidents." A pox on them. Debris, hair, entrails, and feces on their heads.

Summing up: Do not brake excessively or swerve wildly for a small creature, no matter how cute. Do swerve and brake and run off the road for a camel on an empty desert highway, because there's nothing to run into but sand. Never speed in moose country. About deer, I don't know what to tell you. The IIHS study suggests you should brake or swerve only when there's the space to do so safely, never to the point of skidding or losing control, because deer impacts don't reliably injure parties other than deer. The alternative is what, plowing into them? Who does this? People brake. And if they brake hard, the nose of their car drops down and the impact happens lower, perhaps at the level of the deer's legs, causing more of the torso to pinwheel toward the windshield. And the cars behind to rear-end you. What's a rational person do?

Let's ask the most rational driver of all: the autonomous car. If it slams its brakes, does it only do so when no one's tailgating? If it swerves, does it do so only if the path is clear? If either criterion is missing, will it go ahead and run straight over a beagle or a skunk? I posed these questions to Google/Waymo's self-driving car media relations person, but she refused to have relations with me. I got no answers and no one to interview, and soon she stopped replying altogether. Somewhere in the middle of our standoff, one of Uber's autonomous vehicles, traveling 43 mph, plowed into an Arizona pedestrian without braking or swerving. As if she were a squirrel. Seems like they don't have the answers, either.

In 2012, a North Dakota woman named Donna called in to a morning talk-radio program hoping to draw attention to a situation that had been bothering her. She'd been in three car crashes involving deer, and each time, it had happened near a DEER XING sign on a busy road. "Why," she lamented in a recorded encounter that would eventually top one million internet hits, "are we encouraging deer to cross the road in such high-traffic areas?" A short silence followed. "You seem to think," one of the hosts began tentatively, "that deer-crossing signs are telling deer where to cross?" As nicely as possible, he explained that the signs are meant for us, the drivers, to tell us to slow down.

They might as well be talking to deer. Drivers don't slow down when they see a DEER XING sign. This is true of Uniform Traffic Control Device W11-3, the standard black-on-yellow, diamond-shaped sign, and it's true of fancy blinking neon deer warning signs — including the model with three neon deer "activated in sequence to give the impression" of a bounding buck. I am familiar with this technology from a neon sign in front of a strip joint across the street from a San Francisco pizzeria I used to eat at. A bosomy nude danced a sequence of three moves, over and over. I bet you drivers slowed for her.

Warning signs work slightly better when drivers can see some evidence of real danger. This was demonstrated by a researcher who, every Tuesday at dusk, for an unnamed number of weeks, hauled a deer carcass into place a few feet away from a flashing deer-crossing warning. Cars slowed by an average of 7 mph. Likewise, a "realistic taxidermist deer mount" set up in the brush just back from the road, near a blinking neon sign proclaiming "Deer on Road When Lights Are Flashing," caused a 12-mph drop in the average speed.

Did drivers slow down to heed the danger posed by what they took to be a real deer, or did they slow down to gawk at a carcass or, more arrestingly, the odd spectacle of a taxidermy museum specimen in the weeds? I'd guess the latter. That's what drivers do: they ease off the gas and rubberneck. Who wouldn't slow to check out a stuffed deer? Maybe throw it in the back of the truck? *Pull over, Jeb, somebody left a full-body deer mount by the side of the road.* Now you need JEB CROSSING signs.

Nonetheless, a valid concept: pair the warning device with the danger, at least intermittently. Don't cry deer unless deer are around. The best results have come from systems in which the warning message is activated by a signal from solar-powered animal sensors. For instance, a sign that only lights up when something more or less the height of a deer breaks a radar or laser or microwave beam near the road.

So why are we not seeing these systems everywhere? The answer perhaps lies in the woe-laden tale of the Western Transportation Institute at Montana State University. In 2005, they tried out one such system in Yellowstone National Park. Alas, the contractors were unaware of certain laws and park regulations pertaining to what could and couldn't be put up on the side of the road. The system called for erecting eighteen poles per mile, which was far too many, and, PS, they had to be wood, not metal, and they couldn't be that garish color. There were software problems, faulty hardware, budget troubles, topsoil issues, signal failures, false negatives, false positives. Lots of false positives. Drivers were warned a deer was lurking near the road when really it was snow falling. Or a car pulling onto the shoulder. Or moving plants. *Caution! Milkweed swaying in wind!*

When the danger is strictly to the animal, not the driver or the car, forget about it. In 2009, the National Park Service decided to see whether they could save endangered tortoises by installing flashing lights and WATCH FOR TORTOISE signs along two stretches of highway in Mojave National Preserve. A short distance along, they placed a model of a desert tortoise on the side of the road. Then they hid nearby to see if drivers slowed or braked or even craned their necks in a way that suggested they were, in fact, watching for tortoises. Nope.

If only it were possible, as Donna believed, to communicate with the animals rather than the humans. With deer, this has been attempted in various disappointing ways: bumper-mounted deer whistles that don't work, roadside deer repellents that don't work. What has sometimes seemed to help is to mount angled reflectors along the highway to redirect the beam from approaching headlights and enhance "ungulate awareness" of an approaching car. A study in Wyoming recently delivered some promising results, though not with the reflectors they were testing. It was the study's control — reflectors wrapped in white canvas to keep them from reflecting — that worked. The researchers speculate that the deer might have been responding to the white bags as they would to the white butt and underside of a fellow deer's tail. Tail flagging — raising the tail and flashing the white — is thought to be what a startled white-tailed deer does to communicate danger to its companions.

I am skeptical, only because I have read the 1978 paper by researchers at Pennsylvania State University who tried to warn away white-tailed deer by erecting roadside plywood cutouts of deer rear ends with tails a-flagging. On some, the raised tail was painted white; on others, an actual deer tail had been nailed in place. Sadly, because who wouldn't want to see our nation's highways lined with plywood deer asses with decomposing tails, none of it worked.

The Wyoming group also theorized — and I'm with them here — that the reason the death rates were lower with the canvas bags is that drivers were slowing down to look at them. *Hold up, Jeb. White thing on a stick.*

Getting back to Donna for a moment. You could in fact direct deer to where you want them to cross the road — not with a DEER XING sign but with a wildlife-specific over-pass. If the overpass is combined with fenc-ing along the side of the roadway, the animals can essentially be funneled to the safe crossing point. The challenge is that crossing structures and fencing are costly, and it may be impossible to install them in the places they're needed. And with white-tailed deer in some parts of the country, at some times of the year, they're sort of needed everywhere. (Wildlife overpasses and underpasses tend to be used for species that migrate en masse to breed or find food or for populations whose genetic health is threatened by a highway that fragments their population.)

Here's the problem with deer and head-lights. When it's dark, deer don't necessarily make the connection between the little lights and the big car behind. Even if they do, a pair of automotive headlights com-municates practically no useful information as it hurtles closer. Pinpoints of light don't "loom" perceptibly, so it's not obvious that they're approaching, let alone how quickly. Travis and Tom have been testing something that could help: a rear-facing light bar that illuminates the vehicle's grill. The hope is that deer, now able to make out that there is in fact a large object barreling toward them, will get out of the way more quickly and dependably. Now that it's dark, we're headed out onto the Plum Brook back roads to collect some data. The idea is to compare flight initiation distance — FID — and the likelihood of freezing behavior (deer in the headlights) with and without the helpful lighting.

The main Plum Brook parking lot faces a field of corn stubble, that dry, dead snow-dusted brown that is, to me, the color of an Ohio winter. Tom is squatting on his heels, attaching the rig to the front of Travis's truck. He looks up. "Do you hear the woodcock?" The buzzing I assumed was an insect is in fact part of the courtship dance of the male woodcock. In keeping with that rural paradox I will never quite grasp, Tom has a fondness both for wildlife and for hunting it.

While the men ready the equipment, I take their range finder and walk off toward a clique of deer in the field. A range finder is a sort of laser tape measure. Aim the beam onto an object, and the readout tells precisely how far away it is. The FID of these deer, by my estimation, is about three hundred feet. They're wary, with reason. Range finders around here are used mostly by hunters, to calculate bullet-drop compensation when taking aim at distant targets.

Travis yells that we're heading out. I leave off harassing the deer and head back to the truck.

The back roads have few streetlights, making it hard to see a deer in the distance. We're feeling them instead, with a heat-sensing system called FLIR (forward-looking infrared). FLIR displays the world by relative temperature, on a dashboard-mounted display. The image is rendered in a grainy gray scale, like a charcoal drawing. Snowbanks are black. Skunks and racoons in the roadside brambles glow an eerie white. They look like they're made of that strange filament in old camping lanterns. On hot summer nights, a mammal's body heat may match the heat of the asphalt, and Travis and Tom have lost sight of deer standing in the middle of the road twenty feet off.

"There . . ." Tom points into the murk. On the FLIR screen, our first test subject stands in the rough off the righthand side of the road, a couple hundred feet ahead of us. Travis accelerates to a steady 37 mph. Tom's arm hangs out the passenger window and his eyes are on the FLIR screen. He's holding a small bag of gravel covered in reflective tape. As the deer turns back toward the woods, Tom drops the bag. A few seconds later, the truck arrives at the place where the deer took flight and we stop. Tom gets out with the range finder and waits for Travis and me to back up to the gravel bag Tom dropped a moment before. Then he aims the range finder at us. The number it gives him is the FID for this deer.

I'm going to fast-forward here, because I want to share the results of Travis and Tom's project. Their light rig worked, a patent is pending, and the NWRC is seeking a partner to build and market the technology.

Although the deer's flight initiation distance wasn't significantly different, adding the rear-facing illumination caused a dramatic decrease in the incidence of "deer immobility." With the rig in place, only one deer froze in place, as compared with the ten deer that froze in the headlights of the same truck without it.

Back inside the cab, Tom fills out a data sheet. I watch the FLIR screen. A ghost coyote moves through the trees. It stops, looks over at us, continues on its way. A security gate in the distance appears to be manned by a yeti. To see by heat is to understand something key to this work: There are other ways to perceive the world. If you want to communicate something to an animal, you may need to translate your message.

For instance: the light that Travis and Tom are shining on the front of the truck has a strong blue and ultraviolet emission, because that's the part of the spectrum across which deer see best — far better than we can. Deer vision was explained to me in some detail by a man who has studied it: Bradley Cohen, an assistant professor of wildlife biology at Tennessee Tech University. "At dusk, when deer are most active, UV light is the most available thing," he said. While we struggle to see detail in a purply dark dusk, a deer sees it clearly in a bright blue. Our twilight is their noon.

Laundry detergents that claim to give you "whiter whites and brighter brights" provide a service to deer. The manufacturers of these soaps have added an optical brightening agent that emits in the ultraviolet range. To us, it adds no detectable color. To a deer, the clothes are, Cohen says, "luminous." Joke's on you, hunter who washed his camo pants in Cheer.

Whatever advantages a deer has on the ultraviolet end of the spectrum, it lacks at the other end. To a deer, red and orange are perceived as an absence of color. Orange is the new black. So while a safety orange/red hunting jacket makes the hunter stand out to other hunters, to deer it may be more camo than store-bought camo.

Here's something else unusual about deer vision. Deer see best in an elongated strip across their field of vision — a "visual streak" — rather than in a central core of visual acuity, as we do. It would be like reading a book with your peripheral vision. Deer can't read, obviously, but the setup helps them detect a predator trying to sneak up on them.

Some birds have a visual streak, too. It's

useful for hunting prey and for traveling. A migrating bird with a visual streak can study the whole horizon without moving its eyes or head.

Tom Seamans used to do a lot of bird research. He didn't study their field of vision so much as put strange things into it.

11
TO SCARE A THIEF:
THE ESOTERIC ART OF
THE FRIGHTENING DEVICE

Tom Seamans has worked at Plum Brook Station for thirty-one years. He has hair the color of a white-tailed's rump and a soft-spoken, folksy manner. I didn't take a photograph of him while I was here, and my brain, in coming months, will go ahead and substitute Orville Redenbacher in a quilted tan hunting jacket. Tom is a Cornell-educated wildlife biologist and a born tinkerer. He and Travis share a workshop where, you can tell, they pass many contented hours. I've been poking around while they put away the light bar and close up shop. In many ways, it's a typical workshop, and in a couple, it isn't.

The raccoon urine is one way it isn't. Tom and a colleague have been testing potential starling repellents. The urine will go into empty prescription drug vials with perforated caps that are then attached to the floor of starling nest boxes, like giant nightmare

Air Wick dispensers.* Starlings are cavity nesters, and a jet engine cowling is, up until the tragedy of ignition, a cozy postmodern bungalow. Starlings can build a nest in under two hours, potentially setting up housekeeping in the interlude between preflight inspection and takeoff. Bad news for the tenant, bad news for the landlord.

Bird deterrence research has a lengthy history here at the Plum Brook branch of the National Wildlife Research Center. Up through 1990, the focus was not on keeping wildlife away from aircraft and cars but on keeping it away from farmers' crops. When Seamans was hired, in 1987, the work was all about blackbirds and corn.† Tom has

---

* To be sure it was the predator-suggestiveness of the piss smell that was potentially repelling the birds, rather than its foreignness, the team also tested vials containing something just foreign: Febreze Extra Strength Fabric Refresher. Science is here to tell you that starlings feel the same way about Febreze Extra Strength Fabric Refresher as they do about raccoon piss. Neither smell made a bit of difference.

† The NWRC's Sandusky branch was formed through the lobbying efforts of a group of corn farmers who incorporated themselves in 1965 as the Bye Bye Blackbird Association. The name did

deep knowledge in the ancient art of bird-scaring. He's tested dozens of what are known in the human-wildlife conflict field as "frightening devices." Most are only temporarily frightening. It is easy to scare birds away, but much tougher to keep them away. Creatures habituate. They get used to the sound or sight that once alarmed them. They start to call your bluff.

Least — or most fleetingly — effective is the stationary predator decoy. The internet abounds with photographs of pigeons roosting on great horned owl replicas and Canada geese relaxing in the shade of fiberglass coyotes. The classic cornfield scarecrow may actually *attract* birds, because they start to associate it with food. To a flock of migrating blackbirds, it's the golden arches on the side of the highway, the Bob's Big Boy sign, a reason to pull off for a large, fattening meal.

Back in 1981, the human-wildlife conflict researcher and author Michael Conover tested a series of ultrarealistic raptor decoys.

---

not play as well on Capitol Hill as it had in rural Ohio, and in 1967 the Bye Bye Blackbird Association shelved the snickering wit and became the Ohio Coordinating Committee for the Control of Depredating Birds.

Taxidermied museum mounts of sharp-shinned hawks and goshawks were set up at feeding stations to see how long they'd keep away ten species of smaller birds that the hawks are known to prey on. A limp five to eight hours, was the answer.

For long-lasting fear, birds need to see or hear some consequences. In a follow-up study, Conover extended the useful scariness of an owl decoy by incorporating the sights and/or sounds of an actual starling grab. Some involved taped distress calls, some involved a live performance. Here is one of the takeaways: "Tethering a dead starling to the model was less effective than attaching a live starling." Unless you wish to be tethered to the talons of PETA, this is not a viable approach. Besides, in some species, distress cries serve to attract, not repel, the victim's flock mates — to help or sometimes just to gawk.

Depending on what species one is looking to disperse, a dead bird on its own can be strangely effective. Provided it is properly installed. I quote from the National Wildlife Research Center "Guidelines for Using Effigies to Disperse Nuisance Vulture Roosts." "The posture of the prepared bird should resemble that of a dead bird hung by its feet with one or both wings hanging down in a[n] outstretched manner." I spoke to two of the researchers who wrote the effigy how-to manual, Michael Avery (since retired) and John Humphrey, of the NWRC's Florida Field Station. In a 2002 study, they and a colleague documented the effects of rigging up a vulture carcass or a taxidermied vulture on six different communications towers. Vultures like to roost on these and other open-frame towers, and their slippery, pungent droppings make climbing the structures dangerous and gross for repair crews. Within nine days of hanging an effigy, the ranks of roosting vultures were lower by 93 to 100 percent. The birds stayed away as long as five months after the effigy was taken down (or rotted).

Avery was a convert. "Works like a charm."

As with a charm, there is no rational, non-voodooey explanation for why it works. "The answer I give people who ask," Humphrey told me, "is, 'We don't know, but if I went into a neighborhood and saw a person hanging upside down from a tree, I'd leave, too.'"

Avery agrees that, as he and his colleagues wrote in 2002, "It is tempting to speculate that the vultures recognize the taxidermic effigy as a dead roostmate and, not wanting to befall a similar fate, leave the area." He resists that temptation. "This is a fanciful, anthropomorphic notion."

True, admitted Humphrey, "it's not the best answer. But it's the only one I have."

Tom Seamans has a vulture effigy over in a dry-storage room in the workshop, and we are going over there now to visit it. It's one of the models used by USDA Wildlife Services in states with lots of vulture complaints. The body is Styrofoam, because Styrofoam lasts longer than bird, but the wings and tail fan are real, because feathers seem to be the key element.

Given that vulture roost dispersal strategies include shooting at them, the vulture effigy is a lovely if macabre development, and we have Tom Seamans to thank for it.

Like so many discoveries, this one was accidental. Tom straightens a shelf of birdseed bags. "There used to be a big rocket tower here, and the vultures hung out on it." Back in 1999, Tom needed to calculate the average body density of twelve problematic bird-strike species. (He was helping design a standardized bird dummy for testing jet parts.) He headed out to the rocket gantry with his gun. As the dead vulture fell, one of its legs got hung up on the structure, about two hundred feet up. "And I wasn't climbing out on that gantry to get a dead bird." So it was left. No one ever saw another vulture on the tower.

Tom wondered whether the effect could be replicated and put to use. First he tried just laying a carcass out. Nothing doing. It had to be hanging and spinning.

He doesn't know why it works, either. "My only guess is, it's just so unnatural. They're thinking, *Something is wrong here.*"

We can never know whether vultures think such thoughts, but humans assuredly do. The staff at Royal Palm Visitor Center in Everglades National Park tried using effigies to discourage the black vultures that fly over from a nearby roost and vandalize cars in the parking lot. Visitors would return from a day of fishing to find the rubber blades torn off their windshield wipers or the seals around their sunroofs peeled off. Effigies were strung in the trees around the parking lot, which discouraged vultures, but now the rangers spent large portions of their day explaining the effigies to weirded-out park visitors. Now, instead, there's a box in the parking lot with a sign: "Use Tarps to Protect Your Vehicle from Vultures."

Why do vultures do this? Do rubber, caulk, and vinyl off-gas a chemical that's also present in decomposing carrion? Is there one attractive compound common to all these items? Researchers with the NWRC's Florida Field Station tried to figure it out. Because if they knew what that compound — or family of compounds — was, they could use it to lure the vultures elsewhere, in the same way setting up a scratching post lures a cat away from the furniture.

Thus began a project that, from the description in the journal paper, made it seem as though the staid and dedicated scientists of the National Wildlife Research Center had taken to cooking crack in the lab: twenty-one vulture-damaged objects were "finely chopped using a razor blade" and heated to 131 degrees Fahrenheit. The vapors were trapped and then identified by gas chromatography. The idea was to find compounds common to all the materials, and then soak sponges with them and present them to vultures to see how they respond. Alas, the chemicals evaporated too quickly, and then the funding did too. The mystery — and the vandalism — endures.

It's also possible the vulture vandalism has nothing to do with the smell of the materials. My favorite explanation for the behavior comes from retired raptor conservation biologist Keith Bildstein. Bildstein observed the same kind of pulling and ripping behavior in striated caracaras in the Falkland Islands, and he had heard that keas, New Zealand mountain parrots that eat, among other things, dead sheep, also vandalize parked cars in this way. Bildstein noted that the movements and neck strength needed to tear hunks of meat off a carcass were similar to what a bird would need to pull away rubber and caulking. And the stretch and density of carrion muscle and tendon are similar to that of rubber and seals. He speculated that vultures who strengthen their necks by pulling on whatever muscly-tendony items are at hand (or beak) will have a competitive advantage during the avian scrum of a group carcass feed. In other words, it's a fitness regimen.

Tom lays the effigy back down. He does this almost tenderly. He likes turkey vultures almost as much as he likes woodcocks. "Sometimes," he told me earlier, "I'll just lay out in a lawn chair, tip back, and watch them circling around. If it's still, you can hear them sweep by."

Tom shuts off the workshop lights and locks the door. "Effigies," he says, "were one of the things we used at the 9/11 recovery site."

■ ■ ■ ■

When the Twin Towers collapsed, close to a billion pieces of debris had to be searched for human remains. It was the largest forensic investigation in United States history. A thousand people from twenty-four agencies ultimately recovered twenty thousand pieces of human remains. "It was a process involving garden rakes and people literally on hands and knees," wrote Seamans and his colleagues in a 2004 paper describing the process. A large space, remote but not inconvenient, was needed for the operation. A recently closed Staten Island landfill with the awkwardly appropriate name Fresh Kills was selected.

On the third day, the birds came. "We knew what piles were rich in body parts by the way [they] descended on it," said the NYPD inspector who served as incident commander for the operation. "And you would have to fight [them] for the human remains." Seamans was part of the Wildlife Services team brought on to keep the birds away. I've asked him to tell me the story. We're sitting on a beige couch in an old FEMA trailer he and Travis use for watching and analyzing research video footage.

The furnishings are low-budget generic. A human nest box.

"We would get there before sunrise," Tom recalls. "We'd try to keep them away before they landed. We'd walk the perimeter, scanning the sky for approaching birds, and then shoot off pyrotechnics to keep them from landing." Pyrotechnics are harmless explosives that sound like shotgun fire. When the birds habituated to the shotgun sounds, the team got out the actual guns and added some "lethal reinforcement."* One bird is

* My favorite lethal reinforcement story involves Scarey Man, a scarecrow version of those floppy fan-inflated tubular attention-getters you see at strip malls. Unlike those things, Scarey Man also screams and is inflated only intermittently, bursting up like a jack-in-the-box. Birds start to habituate to Scarey Man in about a week, a 1991 USDA study found. Two of the researchers, Allen Stickley and Junior King, then tested whether lethal reinforcement could make Scarey Man scary longer. They'd dress up in a vinyl poncho and sit unmoving on the shore of a cormorant-plagued fish farm. When Scarey Man went off, the man would leap up alongside, "emit a high-pitched wail and bob up and down," and then fire a shot at the cormorants. The log of hours spent "impersonating Scarey Man" resides in the NWRC archives,

shot, and the others take note. It is unpretty but effective. "The only time we would shoot was when they'd gotten to the point where they were completely ignoring the pyrotechnics." Of the thousands of birds run off during the ten-month operation, just twenty-three were killed.

Those twenty-three were used to fashion effigies. These worked pretty well at loafing sites — places where birds hang out together, resting and digesting — but not so much on the recovery site itself. "Because . . ." Tom thinks about how to put this. "The motivation was too high."

I say something about vultures and their ghoulish relish for things dead.

"Vultures?" Tom says. "We only saw one vulture the whole time."

I have made a species-ist assumption. The birds that came to eat the dead were her-

---

along with Stickley's field notes. The birds seemed more entertained than scared. "March 1, 1992. 1456 hrs: Three birds came in and sat and watched me do my thing." Junior got bored and began wandering away from his post and shooting random birds. "I reminded him that the object is to make birds think Scarey Man is shooting," Stickley wrote, disgustedly. What is it with guys called Junior?

ring gulls. Of course! Gulls and landfills. An estimated 100,000 gulls scavenged at Fresh Kills when it was an active landfill.

By chance, my next research jaunt is to a landfill outside San Jose, California, where the inventor of a robotic falcon will be demonstrating his product.

Falcons routinely prey on pigeons and other birds that size, but a gull is about as big as a peregrine and not lightly messed with. It's hard to imagine a falcon taking one on. Or a gull being afraid of a falcon, or a falcon robot. All will shortly be made clear.

For now, RoBird rests in an aluminum case in the trunk of Nico Nijenhuis's rental car. Nijenhuis is the creator of RoBird. He has traveled from the Netherlands, where his company, Clear Flight Solutions, is based, to give demonstrations to potential customers. But first, a slide presentation. We're sitting around a conference table in the Guadalupe Landfill administrative building. Between Nico and some other Clear Flight people is a man from Aerium, a company that provides drone-based bird-scaring services, including RoBird. I'm down at the other end of the table with the people in the Day-Glo vests: the landfill operations manager, its director of engineering, and a man whose card I did not get but who will later tell me, "I've been in waste for twelve years."

Listening to Nico, it doesn't take long to see why any bird, of just about any size, would be inclined to steer clear of a falcon. Peregrines dive at their prey going 200 mph. They, not cheetahs, are the fastest animals on earth. When a falcon arrives at its target, it does what it has come for swiftly and efficiently. "Direct stomp, clean kill," it says in my notes. And yes, they will take on a full-grown gull.

The downside of all that pursuing prowess is that falcons do not excel at the leisurely glide. Hawks and eagles and other "long wings," as Nico calls them, have the surface area needed to coast and ride thermals, hunting as they go.

Nico says something that kind of floors me. "Birds don't like to fly." He means the hectic up-tempo flapping variety of flight. Because it's tiring. Falcons hunt airborne for only five or six minutes at a time, and then they rest. This explains — or is an excuse for — the twelve-minute battery life of RoBird.

"Twelve minutes?" someone says. It might be me.

"Longer would be unnatural," Nico says.

A colleague jumps in. "Let's go see what RoBird can do."

We walk out to the parking lot, where landfill staff are handing out hard hats for us to wear when we get out to the "tipping face." This is where truckloads of garbage are tipped into the "filling canyon" and then compacted and, at the close of the day, covered with construction debris to discourage feral pigs and other nighttime scavengers. The freshly tipped loads are what the gulls are watching for. Even up here by the administrative building, a few are always drifting overhead. Their shadows slide across the pavement.

I introduce myself to the operations manager, Neil. Gulls rank low among Neil's problems. Far higher on the list are the realtors who work the fancy neighborhoods that surround the landfill and tell prospective buyers it will be closing in two years. The homeowners soon realize this was a lie, and they're angry. They've been trying to shut the place down. They complain about the smell, and they complain about the pigs, which wander onto their lawns at night to dig up grubs. The pigs are a holdover from the days when dumps would employ them for garbage disposal.

In the middle of our conversation, Neil turns without explanation and walks off.

"You said 'dump,' " someone offers.

"This is a sanitary landfill," someone else adds. "It isn't a dump."

"*Dump* is a four-letter word."

I guess I'll be driving over with the Ro-Bird people.

A few minutes later, we stand on the lip of the man-made trash canyon. Below, a worker drives a compactor inside a blizzard of western gulls. Like driving anything in a blizzard, there's an element of heightened risk. The gulls make it harder for the drivers to see what they're doing. They also drop things, sometimes surprisingly large things. The hard hats are not just for bird splat.

RoBird is removed from her case. Her exterior is realistically airbrush-painted and textured to approximate the aerodynamic contributions of feathers. Nico opens her cranium to show us where a small compass sits. Then he closes it back up and hands her over carefully, as one would an infant, to a young man whose shirt says "Pilot." The pilot draws RoBird back over his shoulder like a paper airplane and launches her while a second pilot, Ekbert, works the wing controls: a console with two simultaneously operated joysticks, throttle on the left and altitude on the right.

Nico has created something quite astounding here. This is an unmanned aircraft system — a drone — that has no rotors and makes no sound. It soars and dives as a falcon does, on the power of flap, lift, and gravity. The pilot doesn't just steer it around in circles; he mimics the moves of a hunting falcon. All RoBird pilots train with falconers.

If your company buys a RoBird, who will do this for you? No one will, because you can't buy a RoBird. You are buying the services of an Ekbert. A pilot will come out with a RoBird, as often as you need or can afford.

Alternately, you could hire an actual falcon and falconer. Bird-abatement falconers are a thing. They train for two to five years and are licensed by states or, in the case of airport falconers, the Federal Aviation Administration. The San Francisco Giants baseball team looked into hiring a falconer to deter the hot dog–crazed flock of gulls that circles the stadium in the ninth inning, defecating on fans and every now and then dropping down to interrupt play. Nico recommends that even RoBird clients bring in a falconer from time to time for some lethal reinforcement.

Out on the tipping face, gulls are moving off. The usual noisy swagger seems muted. I've seen birds take off when someone fires a pyrotechnics pistol, as Neil did here a few minutes ago, and this is different. "Pyro" causes a sudden mass liftoff, but hang on for a couple minutes, because the birds will be back. It's more of a startle reaction than the kind of lasting low-grade, let's-get-out-of-here nerves brought on by the presence of a predator (or a convincing robotic imitation). Firing off pyro amounts to, as someone here just put it, "exercising the birds."

And irking the neighbors. People don't want to hear fireworks exploding all day. And they don't want to clean droppings off their cars because twenty times a day a flock of gulls circles over their street from the local landfill. They complain, and then Neil has to hand out car wash coupons.

The landfill's recent approach to the gulls has been to ignore them. "It's just less of a hassle to keep them here," said Neil, back when he was still talking to me. Neil is done thinking about gulls, but I'm not, not yet.

12
THE GULLS OF ST. PETER'S: THE VATICAN TRIES A LASER

If you make a gull sufficiently nervous, it will vomit.* While unpleasant for biologists who need to handle them, the habit affords easy insight into what the birds have been eating. Here is a partial list of what a herring gull considers edible — that is, things that have been vomited in the general direction of Julie C. Ellis, senior research investigator at the University of Pennsylvania: bologna, ants, strawberry shortcake, a large mackerel, a whole hot dog, intact mice, squid, a used sanitary napkin, discarded lobster bait, Vienna sausages, an eider

---

* When I first heard about "defensive vomiting," I figured it was a way to become lighter and better able to take wing and flee. Nope. Nor is it done to repulse the predator. *Au contraire,* it's more likely, said gull expert Julie Ellis, "a way to distract a potential predator with some alternative food." Animals are different from us.

duckling, beetles, rotten chicken drumsticks, a rat, a paper muffin wrapper, a loaded diaper, and a plate's worth of spaghetti marinara with mussels.

No gull has upchucked flowers at Julie Ellis. Gulls would eat the world but leave the plants. So when Dutch master florist Paul Deckers arranged three truckloads of blooms around the altar outside St. Peter's Basilica on the eve of the pope's 2017 Easter Mass, he did not worry about gulls.

And yet. "What we saw was not to believe." When he'd left St. Peter's Square the evening before, six thousand daffodils lined the wide, shallow steps to the outdoor altar. He returned at 6:00 a.m., mere hours before the crowd would be let into the square, to floral mayhem. Potted daffodils lying on their sides in the center aisle and helter-skelter on the steps. Potting soil on the chancel floor. Long-stemmed roses had been plucked from vases and strewn about, as though a diva ballerina had come through for a farewell performance sometime in the night.

Yet the blossoms weren't eaten. It appeared to be an act of senseless vandalism, as much a mystery as the vinyl-pulling vultures of the Everglades or Dipanjan Na-ha's cooker-smashing macaque. Why would

319

a gull do this? Was there a biological motive? Are some species just dicks?

For answers I turned, via online meeting software, to Julie Ellis at her desk in the laundry room of her home. "The only thing I could imagine," she said, "is that they were looking for worms in the soil." Could be. But most of the ravaged daffodils I could see in the pictures Deckers sent were still in their pots. And the roses were cut flowers.

Fellow gull researcher Sarah Courchesne had joined the conversation, from her car in a parking lot in Maine. Sarah studies herring and great black-backed gulls at Shoals Marine Laboratory on Maine's Appledore Island. "What do you think, Sarah?" said Julie. "Sarah? Looks like she's trying to get the sound to work."

Sarah's mouth moved silently and then her face froze and then at last words came through. "Well, that was a tense moment. Was it a form of grass-pulling, Julie?"

"I thought of that," Julie said. "Maybe." In breeding colonies, gulls use their beaks to pull up tufts of grass as a territorial display. "Displaced aggression" is the term Sarah used. Like punching the wall instead of someone's face? Yes, Julie said, though gulls do the latter sort of thing, too. At breeding colonies, herring gulls will sometimes peck to death another gull's chick should it blunder into their territory. And then they, or another gull, may eat it. I read all about this in "Cannibalism in Herring Gulls," an article by Jasper Parsons, who watched a lot of it go down on Scotland's Isle of May. I showed off my new vocabulary word: *kronism,* the eating of one's offspring. They also do that.

Based on Sarah's observations, reports of herring gull cannibalism are overblown. "There are sizable numbers of times when they kill the neighbor's baby but don't eat it. Or they don't even kill it. Just peck at it."

"You know," Julie chimed in, "maybe blind the neighbor's baby, leave it lying there as it slowly dies. It's lovely, really."

Still, neither Julie nor Sarah would say that gulls, as a species, are dickish.* While 20 to 30 percent of herring gull chicks that wander away from their nest are attacked, a similar number, one study showed, are adopted by a neighbor who feeds and

---

* Especially as they don't have dicks. Like most birds, gulls mate by aligning their cloacal openings. The ornithological term for this is "the cloacal kiss." Which makes bird sex sound sweet and demure, until you remember that they also excrete through their cloaca.

protects them. As with humans, as with bears, a few individuals are responsible for the bulk of the species' churlish (to us) behavior. Of the 329 herring gull chicks cannibalized during the 1968 Isle of May breeding season, 167 were eaten by just four cannibals. According to Parsons, one out of 250 herring gull pairs practices cannibalism. It has nothing to do with food shortages, he found. It just seemed to be what they liked to eat best.

Gulls have evolved a generalist's bill and a thick gizzard that withstands and regurgitates shell fragments (and baby gull beaks and diaper linings), so they can pretty much eat what they want. Individual gulls are as different as individual humans. Some keep to the shore and fish for their living. Some hit the landfills. Still others commute into the cities, where they eat pavement fries and hot dog leavings (and, one study suggests, get coronary artery disease). A few eat their neighbors' children, and a few are hunters, nabbing flickers and songbirds on the wing. Sarah recently saw a great crested flycatcher in someone's vomit. "Although," she added, "having been through the esophageal chute both forward and in reverse, it was more of an *okay* crested flycatcher."

There is, or there was, a hunter gull that

hung around St. Peter's Square, site of the aforementioned floral vandalism. We know this because the bird was caught on camera in 2014. You can watch it in slow-motion as it swoops in, beak first and irony ablaze, to nail the white "peace dove" that Pope Francis had just released. Every January the pope appears on a balcony with children from a Catholic youth group to read a message of peace and release a dove. The dove survived, but the tradition did not. In later years, a helium-filled balloon in the shape of a dove was released.

Julie and Sarah wish you to know that gulls have a "more endearing side," which we would see if we spent time in their breeding colonies. Gulls are devoted parents, and that includes the males, who stick around to help raise the chicks. This is not typical for birds. More typical is the starling, whose nest behaviors were observed a century ago by naturalist F. H. Herrick and described in his *Home Life of Wild Birds:* "In the space of four hours . . . [t]he female brooded her young over an hour, fed them twenty-nine times, and cleaned the nest thirteen times. The male made eleven visits" — whether he fed his chicks or just sat around and got in the way is not made clear — "attending to sanitary matters but twice."

Gulls are community-oriented, provided the community is their own. When a gull sights what it perceives to be a predator, it warns the rest of the colony. The frequency of its alarm calls communicates the intruder's approximate location, and others rush to harass it.* This of course contributes to their dickish reputations, because the intruder, in coastal resort towns, is often a tourist who has unknowingly wandered too close to a nest. "Pensioner Left with Blood Pouring Down Her Face After Seagull Attack," says the British ITV website, sensationalism pouring from its headline writer. (The accompanying photograph shows only a small patch of dried blood on the crown of the woman's head.)

Where India has monkeys, it seems, we have gulls: bloodying old ladies, stealing snacks from tourists' hands, boosting newspaper sales, vexing politicians. ("Seagull At-

---

* And good luck if it's you. In addition to beaking your head, gulls can be, as Ellis phrased it, "very adept at aiming their feces." She shared the story of a student on Appledore Island who, hoping to protect herself while traversing a nest-dense canyon, put on a raincoat and pulled the hood up tight. "A gull managed to shit directly into her mouth."

tacks: David Cameron Calls for 'Big Conversation' About Issue" was the *Guardian* headline after a gull stormed a pet tortoise in Cornwall.) Gull researcher and College of the Atlantic professor John Anderson sees gull "attack" media hype all the time in the coastal town where he lives. "It's absurd," he said to me in an email. "Dogs barking and lunging at people are so NOT news, whereas a gull diving at you gets press." Anderson throws some blame on *The Birds*. "Alfred Hitchcock has a lot to answer for."

So let's focus on gulls' endearing qualities. I recently read a book about gulls by a former *Audubon* magazine editor. The book mentioned, and I then mentioned to Julie and Sarah, that gulls have a food-sharing call. How lovely is that?

"Hmm," said Sarah. "The call I hear them make when they find food is the long call, which is territorial. I would bet they are staking a claim to the food they find, rather than inviting friends to dinner. So I must plant a flag in the gulls-are-dicks hypothesis on this one."

But that's a survival strategy. All of this is. It's about keeping fed, protecting one's progeny, escorting the genes to the next generation. It's gulls being gulls and, unfortunately in some cases, trying to do it too close to people being people.

But messing with the pope's roses is . . . who can say what that is? I better go see. Easter weekend is coming up, and this time the Vatican is prepared. Along with Paul Deckers and his team, they've flown in a "laser-operated scarecrow" and its creator, André Frijters. I keep hearing about lasers as an effective, seemingly benign bird-scaring method. Tom Seamans used one in the evenings at Fresh Kills. Staff at the Royal Palm Visitor Center tried one with the car-vandalizing vultures. Could keeping birds away be as cheap and easy as pointing to graphs on PowerPoint slides that no one understands?

Easter weekend at the Vatican is a joyful thrum of Catholic professionals. Nuns and priests jet in from all over, giving St. Peter's Square the look of a college campus on commencement day: the flowing robes, the specialty headwear, the selfie sticks. A dozen or so gulls are here, too — cooling down in the fountain, watching the crowds from St. Peter's marble haircut — but for now there's no need for deterrence. Six assistant florists bustle around setting up the flowers. André Frijters is meeting me at the security gate in an hour, around 5:00 p.m. He can't begin his work until all the flowers are in place, because the laser beams must be set to reach every plant. I wander the souvenir stalls and pontifical tailor shops, the low and the high of Vatican shopping. I try, and fail, to talk my way past one of the Vatican Swiss Guards, partly because I'm bored, and partly because a man in ballooning striped knickers doesn't seem all that imposing.

From behind a barricade, I watch Paul Deckers orchestrating the final adjustments. For most of my time here, he is a figure glimpsed at a distance, always in a hurry, striding through my field of vision in leather hiking books and a fanny pack (also leather). The night I arrived, he alit, briefly, to chat as I sat with André at the cafe in their hotel. He shared the story of the 2017 gull debacle and how he'd gone on Dutch television when he got back to the Netherlands to crowd-source a solution for future Vatican gigs.

Of the 250 or so people who responded, most had not thought through the singular exigencies of the scenario. From where I stand, I can practically see inside the pope's bedroom. The residents of Vatican City don't want to be kept up all night by "sirens with yelling sounds," as one person suggested, or "bomb sounds." They don't want to attend Mass the next morning in a fog of "smelly odors." And although an effigy could be seen to complement the Crucifixion iconography already on display at St. Peter's, a dead gull suspended by pinioned feet was likewise not up for consideration. André wasn't even sure an effigy would work at night. Gulls evolved as diurnal birds; though some have begun taking night shifts to raid Roman garbage bins, gull night vision may not be sufficiently acute to process the creepy particulars of an effigy. I wondered about RoBird, but apparently Nico Nijenhuis hadn't been watching TV that day.

André telephoned Deckers the day after the televised plea and said to him: "I have been scaring birds for twenty-five years." He owns the company Vogelverschrikker (Dutch for *scarecrow*). André suggested bringing in a LaserOp Automatic 200, a sort of monochrome laser light show. Lasers are silent, seemingly humane, and they can usually be counted on to unnerve gulls for at least a week. They're mainly used in darkness or low light, when the beams are most visible. Wildlife Services, in the States, has used them to dissuade cormorants, gulls, and vultures from roosting on structures where someone doesn't want them — and their droppings — to gather.

The laser beam is green, a color some birds are thought to see better than we do. There's a theory, set forth on some manufacturers' websites, that the lasers work because these birds perceive the beam as a solid green rod slicing through the air and coming at them. (I wonder whose theory this is and how many *Star Wars* movies this person has seen. I checked: no "lightsaber" references in the product copy. If anything, they call it "the stick effect.") Some species — pigeons, for instance, and starlings — either don't see it this way or aren't fazed.

At five sharp, André shows up and walks with me to the altar. Tomorrow eighty thousand Catholics and tourists will come to hear the pope say Mass and take communion from a roving platoon of priests, but for now we look out on an audience of gray plastic chairs. André has brought two Laser-Op Automatic 200s. One would suffice, but this is, you know, the Vatican. He wants to be sure. The lasers are housed in a boxy white structure that, out in a farm field, might be mistaken for stacked beehives. Here, at the foot of the altar at St. Peter's Square, they look like, I don't know, Ikea baptismal fonts?

André and I attempt conversation over the noise of a cleanup crew with a leaf blower. "How did you get into bird-scaring!" I'm yelling.

He leans in toward the side of my head. "Well, I was a farmer myself!"

The possibility hadn't occurred to me. André looks like a farmer the way Paul Deckers looks like a florist. He, too, is wearing leather — in his case, a lived-in black jacket. His hair and his stubble are buzzed to the same length. His jeans sit low on his hips and while they are not tight, they are for sure not farmer jeans.

André grew lettuce. "Iceberg and little gems!" The leaf blower shuts off in the middle of this. "Little gems!" comes out like a sports cheer.

A salad-loving menagerie went after his lettuces. Hares, crows, wood pigeons. "The wood pigeons would eat the hearts. I couldn't get them away with the gas bangers." (A gas banger, or propane cannon, is a pyrotechnics device left out in a field and set to fire at intervals.) "You can put netting on top of the crops, but it's expensive and a lot of work to keep it up. The crows would pull up the plants because they'd think there's a worm underneath." I ask André whether that might be what the gulls were doing with the daffodils. He says he thinks the damage was done out of curiosity: "The gull thinks, *There is something different, maybe I can eat it. So I try it first.*"

André likes birds. He mentions their helpful consumption of insect pests. "The birds were here before the farmers," he adds. "The farmer comes along and opens a restaurant and the birds are coming to eat, and that's generally how it goes."

Like the senator from chapter 8, André undertook the most effective strategy of all. He changed careers. One day, while he was still a farmer, André was on the phone with the manufacturer of Scarey Man, the noise-making, intermittently inflating pop-up bird deterrent. "He said to me, 'You know, I don't have an importer in Holland. Would you like that?' 'Okay,' I said. 'I like that.' " André became the Scarey Man man for the Netherlands. Customers would often ask for a cheaper product, and soon he began stocking pyrotechnics, hawk kites, effigies. "Everything you can buy in the world about bird-scaring, we had it in the warehouse." For a while he kept up the farm. "Between the scaring and the lettuce, it was too much." He went with the scaring. The hours are better, and the money. As with a gold rush, so with farming: the people who make the most reliable living are the ones who sell the supplies to everyone else.

André gets up to walk the perimeter of the floral display, squatting every few feet to sight the laser and note any points it might miss. I tag along with my notepad, interrupting his focus and bumbling into his sight lines.

"I'm just going to follow you around."

"I see that."

From somewhere beyond the square, a clamor of whoops and cheers coalesces and crescendos. It's a red-carpet sound — the A-list star pulling up and then stepping from the limo. "It's him," André says. "Francis. He's a rock star."

Hundreds of nuns are in town this week. I've seen nuns with green habits, pink habits. Bad habits! Nuns vaping, nuns cutting in line. Right now, nuns are running, laughing, elbows pumping and veils flying. The security gates on the north side of St. Peter's Square just opened, and a group from the front of the line are rushing for front-row seats at whatever services Pope Francis is about to hold inside the basilica. As they race behind the outdoor altar, a gull on a pole-mounted speaker takes flight. Two on the cobblestones follow.

The nuns have just demonstrated the world's oldest technique for scaring birds away from plants: have someone run at them making noise. The best means of keeping away crows and other "enemies of the corn," wrote Gervase Markham in 1631, is "to have ever some young boy . . . to follow the seed-man . . . making a great noise and acclamation." The practice was documented in detail for an 1869 British parliamentary investigation of child labor practices which was the subject of a recent museum exhibit in Nunney, England. The children were usually between six and nine — boys not yet strong enough for more laborious farmwork. They were paid a pittance to roam the fields "hallooing" and clacking wooden bird scarers. They worked a month in spring when the seeds were sown and came back in fall for another month when the crops were ripening, effectively obliterating any continuity in their education. The drawback from the farmers' perspective was that the boys were lazy, to the extent that, by one farmer's words, "each boy required a man to look after him."

Well, halloo, they should have just had that man do the scaring. Adult bird scarers are uncommon, but they do exist, even today. Hiring them can be cost-effective. I say this with some confidence because science has looked into it. The UK Ministry of Agriculture, Fisheries and Food maintains an experimental farm in north Norfolk, England, for testing, among other things, "scaring regimes." The site, a series of coastal wheat and rape-seed fields, was selected for the noteworthy annual depredations of a flock of three thousand resident brent geese. Researchers Juliet Vickery and Ronald Summers used the site to compare the cost-effectiveness of commonly used techniques — propane cannons and the like — with that of a "full-time human bird scarer" tooling around on an ATV six days a week. ("The farmer scared geese on the seventh day," the researchers wrote biblically.) The human bird scarer achieved a significantly greater reduction in the amount of time the birds spent grazing and the intensity — as quantified by "dropping density" — with which they went at it.

Even factoring in the initial outlay for the ATVs, the human scarer was more cost-effective. A few farmers would seem to have taken note. Dutch berry growers sometimes hire college students to work as bird scarers in the summer as the crop ripens.

A human bird scarer makes sense for a relatively small acreage. The flowers outside St. Peter's cover less than half an acre and require just a couple nights of protection. If ever a scenario called out for simple, cost-effective human bird-scaring, here it was. If I had seen Deckers's original call for ideas, I'd have sent an email saying, "Did you think about hiring a human bird scarer? Just some person to sit there at the altar and keep an eye out and run off any gulls?" Instead I sent the email after I got home from Rome. "No, we did not think about that," came Deckers's reply. "It's better for André that we did not!" And it was. The Vatican City State is gearing up to be a nation with one grocery store, one pharmacy, no movie theater, and two LaserOp Automatic 200 bird-scaring units.

André opens the lid of one of his laser units. Inside there's a digital display, like something Daniel Craig would be hunched over, trying to defuse the bomb that's set to take out MI5 and the whole London waterfront, or program his sprinkler system. "Only five buttons," André says. Buttons for setting the start times and the intervals between activations, and buttons for setting the boundaries of what needs to be covered. "A farmer can do this."

For a large farm, using human scarers would either be personnel-heavy or endlessly Whac-A-Mole. An automated system like André's holds great appeal: a series of solar-powered units that can be custom-programmed to cover the whole field. Is this the golden future of bird deterrence?

André drags a planter out of the beam's path. "In five or six years," he begins, but he doesn't go where I'm expecting, "no one will be using lasers. It's dangerous." Even the handheld kind sold for classroom lectures can damage a retina. When laser light is absorbed by pigments in the eye, it deposits energy and heats up the tissue. Because the light arrives in a tight beam — and is further focused by the eye's lens — the energy density is high. In terms of the damage caused, think of the difference between someone in stilettos stepping on your foot and someone wearing loafers.

André isn't wearing laser safety goggles. When asked, he points out that the beams are aimed downward, and that you would need to be looking into the source of the beam to harm your eyes.

Who stares into a laser beam? Adolescent boys, in 80 percent of the cases. That was the finding of a team of ophthalmologists who reviewed 77 cases and emailed surveys to hundreds of their peers. Some of the kids, they found, were taking part in laser staring contests, reluctantly admitting to emergency room personnel that they'd looked into the laser for 10, 20, in one case 60 seconds. (The boys often had behavioral or genetic disorders linked to self-injurious behavior.)

Before I left, I spoke by phone to a Purdue University biology professor and former National Wildlife Research Center researcher, Esteban Fernandez-Juricic. He was involved in a study examining the safety — to birds — of brief exposures to a bird deterrence laser. It was the first study of its kind, previous safety claims having been based on human focal lengths, spectral sensitivities, and eye configuration, which have, as he put it, "zero to do with the bird eye." Some species, Esteban said, don't react to a laser, even when it crosses their visual field. "Maybe some species cannot see the lasers. We don't know. So with these species that don't respond, we should take more care." But the opposite may be more likely to happen. Esteban voiced the thoughts of an imaginary frustrated farmer faced with a flock of one of those species: *"Come on, species, there you go! More to you!"* He has a wonderful, excitable

Roberto-Benigni-wins-an-Oscar way of speaking.

Laser companies know about Esteban's project, and it makes them uneasy. They've been scrambling to manufacture large units for agriculture, and a lot of money is at stake. "I have stories which I cannot tell you," Esteban said. "It's . . . how can I characterize it . . ."

"Someone tried to influence your results?" I offered. "A bribe?"

"What you are saying might not be very far from what happened once." Esteban had to call the university lawyer. "I said, 'Oh my gosh. You need to get involved, because this is . . . ! *Whoa.*'"

André Frijters had not heard of Esteban, which was not surprising, because the results of the study had not yet been published. Three months after I got home, Esteban shared preliminary findings with me — no details, and no laser names. The results, he said, were unsettling enough for one bird-scaring outfit to travel to Purdue for a meet-up.

Esteban tried to calm his visitors. "It's not like, *You bad companies that are trying to do this!* It's an opportunity to work with the scientific community to modify the way the laser is operated, or the wavelength or the intensity."

One year later, the study remained unpublished and Esteban was not replying to my emails. I hope he's okay.

Around 5:00 a.m., about the time of the Easter 2017 flower assault, I walk over to St. Peter's Square to see what's happening. The lasers make firefly flashes as the beams touch the plants in their whizzing circuit of the altar area. They appear to be doing their job. From what I can see from back behind the security fencing, Deckers's flowers are unmolested. Thirty gulls are asleep at the base of a fountain in the center of the square, drawn to the heat of the cobblestones.

Along the colonnade, a dozen of Rome's unhoused are waking and quietly rolling up their sleeping bags. Police cars are parked thirty feet away, but the *polizia* have left these men and women to sleep in peace. Is this the kind and gentle hand of Pope Francis? I wonder to what degree his papacy is influenced by his namesake, St. Francis, friend to the poor and friend to the animals. Might the current papacy espouse more progressive approaches to nuisance wildlife? A ludicrous inquiry, I know. But I'm here. I might as well ask.

13

THE JESUIT AND THE RAT: WILDLIFE MANAGEMENT TIPS FROM THE PONTIFICAL ACADEMY FOR LIFE

The Vatican City State is a sovereign nation the size of Disney's Magic Kingdom. Like Disney World, it has places where tourists may go and places where only staff are allowed. Because I'm neither, and because I don't fit the Vatican's definition of press, I was told to write to the secretary-general of Vatican City State. It was like trying to enter the United States by writing a personal note to Donald Trump. Though likely went better. My note was forwarded, and promptly generated a gracious reply, which Google Translate presented as follows. "I gladly take this opportunity to offer you, my dear lady, the expression of my distinguished homage." Signed, somewhat incongruously, "the Vatican Director of Gardens and Garbage."

The very same director, in a blue Ford Focus,* has just pulled up along the curb at

* Pope Francis rides around in a Focus, too.

one of the guarded entrances to the Vatican. Rafael Tornini gets out to shake my hand. In person, he is a formal but not showy man. He is dressed in a dark blue business suit, clean if slightly worn. We head out for his office. The streets are narrow and largely empty. It appears to be a city without traffic, a country without children.*

"And there is the border with Italy!" I fol-

What's up with the Vatican and the Ford Focus? Nothing, says Erich Merkle, of Ford's marketing department. Coincidence. Merkle drives a Mustang, but he defended the lowly Focus. "Visually, the lines are actually pretty handsome." He added that in Europe, Ford sells a turbo-charged 252-horsepower Focus called the ST. "That thing's a monster. Once you option it up? Put a spoiler on it, some Recaro bucket seats in the front? It's really hot." Was the pope's ride in fact a Focus ST? It was not. The pope opted for the "more rudimentary" model.

* A few high-ranking members of the Swiss Guard live with their families inside the Vatican City State, but no baby has ever been born inside the nation. When journalist Carol Glatz was nine months pregnant and working in the Vatican press hall, a nun working alongside her hoped the baby would be born right there. Because it would have been the first birth ever registered in the Holy

low Tornini's gaze to the massive wall that surrounds the Holy See. A gull glides over. There's your symbol of peace, I think to myself. A bird, any bird, soaring over walls, ignoring borders! Peace, freedom, unity! It's possible I've had too many espressos.

Tornini's office is modest. His view is of the leaves of some thick vine. An interpreter joins us. Tornini says gulls cause no trouble inside the Vatican gardens. "Here it's the green parrots." They eat the seeds that the garden staff plants.

Nothing is done to get rid of them. "They're part of the system." According to Carol Glatz, the Catholic News Service reporter who covered the peace dove fiasco, the Holy Father is a bird guy. He used to own a parrot (and he taught it dirty words).

Pope Francis has indeed steered the nation according to the worldview of St. Francis of Assisi, the OG animal activist. Shortly before Tornini took his current post, Pope Francis decreed that biological pest control be used in place of chemical pesticides. Insects that prey on problematic bugs have been introduced, and nest boxes have been

---

See. What splendorous benefits might befall a Vatican anchor baby? Zippo. Vatican citizenship is granted by administrative decision, not birth.

mounted on tree trunks in the gardens to encourage bats, because the bats eat the mosquitoes.

We get back in Tornini's car to go see the Vatican bat boxes. They're very nice, they're wooden, they're tasteful. Soon we arrive at a low, sprawling mound of grass clippings. The Vatican compost heap! Far in the back, a week old by now, is some distinctly papal organic waste: elaborately woven palm fronds left over from Palm Sunday. Tornini pulls one out for me.

Open compost piles can attract animals. Here it is — that rare conversational segue to the topic of Vatican pest control. I ask the interpreter to inquire about rodent problems.

Ratti, I hear him say to Tornini. He turns to me: *Sì, sì.* The Vatican has rats.*

* After I got home, I went on the web to see if I could find out which pest control firm the Vatican uses. A search brought me to a computer-generated web page for what had to be a non-existent company, Derattizzazione Roma. Under the heading "Our greatest services for Vatican de-ratting" was a crazed two-page list that, were it real, suggested a fabulously pestilent Vatican where deratization was underway day and night and even on Sunday, in every room of every building, where

Sì, answers Tornini when I ask whether they set traps. He says something to the interpreter, who then adds, "They have to do an action against them, because it's a big population. And they really make damage. To the machinery, the wires. They try to keep everything as clean as possible, but —"

"So Pope Francis is okay with killing rats?"

Tornini has never even met Pope Francis, and now he is being asked to speak for him. The interpreter listens, then turns to face me. "He says you should ask him."

Of course I can't do that. Instead, I will do the closest thing available to a lapsed Catholic with no high-level connections. I have an interview with Father Carlo Casalone, a staff bioethicist at the Pontifical Academy for Life. The PAL is a Catholic

---

even the mice had rats. Urgent Vatican Deratization, Night Vatican Deratization, Deratization of Vatican Canteens, Sunday Vatican Deratization, Deratization of the Vatican Shopping Centers, Deratization of Vatican Technical Rooms, Deratization of Vatican Elevator Lifts, Deratization of Vatican Mice. The exuberant translation software had even created a drinking establishment for exhausted deratizers to repair to when their shift was done: Rodent Pub Vatican.

think tank of sorts. Its members are appointed by the pope but are not necessarily clergy or even Catholic. The PAL guides Church doctrine on matters ranging from the perennial (abortion, euthanasia) to the more cutting-edge (gene therapy, artificial intelligence). As the sexual abuse scandal widened, the PAL played a role in drafting the Church's response. I told the PAL media manager I was interested in the academy's opinions on the designation of certain wildlife as pests. That is, under what circumstances should a species be exempt from moral protections against extermination or cruelty? I quoted St. Francis of Assisi. I did not mention rats. He replied right away. It was perhaps a welcome diversion from thornier inquiries lately clogging his inbox.

Walk straight down Via della Conciliazione from St. Peter's Square. Three blocks along, across the street from the souvenir store selling Pope Francis bobbleheads, with their unfortunate suggestion of tremor, you will see a boxy three-story building of caramel-colored stucco. A plaque beside the doorway announces that you have arrived at the Pontificia Academia pro Vita, the Pontifical Academy for Life. Although situated outside the physical borders of the Vatican City State, the PAL is officially a part of it, which means that while visiting, you undergo a sort of geopolitical transubstantiation. You are in Italy, and yet you are inside the Vatican.

Father Carlo's office walls are white and undecorated, unless a crucifix counts as decor. It was the same at Tornini's office. The extravagance of the Vatican seems concentrated, like a laser beam, inside the museum and the churches. Father Carlo is himself unadorned: black pants, black shoes, black button-down shirt with the white tab collar. His voice is low and quiet and he doesn't talk with his hands, or not in the stereotypical manner of the Italian male. Though the floors are marble, I imagine him making no sound as he walks.

Some of this anti-ostentatiousness may be the influence of Pope Francis, who in turn is influenced by St. Francis of Assisi, the humble, nature-besotted Capuchin. When he became pope, Francis took a regular clergy apartment. Like Tornini, he gets around in a Ford Focus. This week I dropped by a Holy Thursday Mass that included the ritual washing of a few congregants' feet. It was more a gesture than a real cleaning — a splash of water over the instep. "Francis gets right in there with a scrub brush," laughed my Catholic News Service acquaintance Carol Glatz. (And yes, he eschews the red loafers.)*

* I almost wrote "Prada," and then I read Dieter Philippi's exhaustive treatise on *campagi,* papal shoes, replete with more than 100 photographs of custom-made red papal footwear. Pope Benedict's official shoemaker was Adriano Stefanelli. He made the red loafers that earned Benedict the *Esquire* honorific "accessorizer of the year," as well as "the special slippers which the pope wears around his apartment." Gammarelli, a papal and clerical tailor shop near the Vatican, has a shoe-maker who by tradition makes the ceremonial red loafers worn by newly elected popes when they first appear on the St. Peter's balcony. And anyone else, sniffs Philippi throughout his 119-page commentary, is lying. Silvano Lattanzi's claim that Benedict wore a pair of his velvet slippers? "I am quite certain this is a false assertion." The mules made for him by Raymond Massaro? "I do not think the Pope has ever worn this kind of shoe." Ferragamo's wine red loafers in "the papal style"? "The Pope has never worn these shoes." The slip-ons imprinted with the Vatican coat of arms? "The Pope would certainly never wear a design of this kind." Those Prada red loafers with the ornamental seams? "A fallacy. The Pope renounces orna-

So I'm curious. How far does the pope think we should go in the direction of respecting and protecting the natural world and its wild inhabitants? Before I arrived, the PAL media manager sent me a copy of Francis's rather beautiful encyclical *On Care for Our Common Home.* "Each creature has its own purpose," he writes. "None is superfluous." He describes how St. Francis would burst into song when he gazed at the sun, the moon, or the smallest of animals. I read these passages to Father Carlo.

He listens, nodding. "Saint Francis began a new relationship between nature and humanity. If you read his poems, you find the expressions Sister Water, Brother Sun, Sister Moon."

"Would Saint Francis include Brother Rat?" Sister Boll Weevil, Uncle Blackbird Who Devours 2 Percent of the North Dakota Sunflower Crop?

Father Carlo says yes, yes he would. "He

---

mental stitching." According to Philippi, only one commercial shoe company can rightly claim to have shod Pope Benedict. While hiking during his summer 2009 holiday, Benedict wore Camper Pelotas leather sneakers.

includes even death."[*]

"Did Saint Francis say anything specifically about rodents?" I hear myself say.

"No, he didn't. But the point is, brotherhood is not a simple relationship. With your brothers and sisters, normally you fight. You cannot think that there is an idyllic way of being in a relationship with someone. Every relationship among humans and the earth is not only connotated with positive aspects. At the same time you also have negative aspects. The point is, how do you deal with those aspects." He's good, this guy.

"Yes, and how *should* we deal?" It's well and good to say these things, but how do we act in a way that serves both human and animal fairly? Let's take the example of Canada geese on golf courses. What is their crime? Befouling the turf.[†] Littering. For

---

[*] Sister Death — it's a woman. Also the title of the seventh album by Alec K. Redfearn and the Eyesores. *Signal to Noise* magazine describes Sister Death as "a gorgeous amalgam of 20th Century Americana, cabaret . . . Eastern European folk, noise rock and minimalism." You never know where a footnote will take you.

[†] But not as heavily as the internet would have you believe. Goose Busters has them extruding 3 to 4 pounds a day. The Geese Police superinten-

this, should we be allowed to call someone in to round them up and gas them? Do they deserve to die because a few well-heeled humans want to hit a ball into a hole and

---

dent at the National Mall in Washington, DC, claims 2 to 3 pounds a day. A Boston city councillor: "as much as 3 pounds a day." The Canada goose fecal smear campaign appears to have hit its zenith in New Jersey's *Montclair Local:* "An adult Canada goose can weigh up to 20 pounds and defecates more than twice its weight daily." That would be *40 pounds a day,* coming out of a single goose. That's how much a *horse* makes. The reporter cited the USDA; a contact at their NWRC public affairs office steered me to the USDA "Geese, Ducks and Coots" fact sheet, which gives a daily total of 1.5 pounds. The author of the USDA fact sheet got his information from a Virginia Tech University Cooperative Extension goose fact sheet. *That* fact sheet says "studies have shown" but does not cite any studies. A Google Scholar search brings up just one researcher, B. A. Manny, who actually went out and weighed some turds. Manny's findings: the average total (wet) weight of a Canada goose's daily droppings was just a third of a pound. Where did Virginia Tech get the 1.5 pounds per day figure? The author did not reply to multiple emails, so it remains a mystery.

they need an obsessively tidy playing surface the size of the Holy See? Think of all the Sister Water that gets wasted watering the greens. Maybe it's time to eliminate golf, not geese!

Father Carlo collects his thoughts. Among them, surely: *Who let her in?* "We have to put the action in the context of where we are. What does it mean, the golf field, for the people that are working in it? If this is the only way that people can find employment in the region, you have to keep in mind also this aspect of the action you are performing.

"Secondly, maybe it's not necessary to kill the birds. You can act in another way to deviate the trajectory. You have to move, to think, in a progressive way of intervention."

Like egg addling! I almost blurt. Rather than culling flocks of Canada geese, some municipalities seek out nests and either shake the eggs or coat them in oil and then return them to the nest. With the result that the parents are incubating blanks. To figure

---

Poundage aside, the Canada goose is a frequent crapper. Twenty-eight times a day, on average, Manny found. In related research, a Canadian team reports that "sleeping geese sometimes produce small piles of droppings."

out the cutoff for humane termination of a Canada goose fetus, a team from the Michigan Department of Natural Resources examined tens of thousands of eggs. They then established a method to assess the age of an egg by seeing if it floats — an indication that there's more air inside than goose. The technique has been recommended by the Humane Society of the United States and by PETA, and while I wonder what the Catholic Church might have to say about goose abortion, I don't wish to addle Father Carlo, so I move on.

What about a predator, say, a coyote, that kills a person's pet? Is it ethical for the person to kill the predator? When the predator acts by instinct, to survive?

Father Carlo aligns a stapler on his desktop. "You have to keep in mind the emotional impact for the people."

"But how do you weigh these things, Father Carlo? The feelings of the person versus the life of the predator?"

There is a knock on the office door, and a man steps in carrying a tray of traditional Italian Easter cake and a carafe of water. "Ah, Sandro, *grazie*!" Father Carlo seems delighted by the refreshments or maybe just the break in the conversation. Sandro sets down the water glasses. Mine has a bit of brown gumpy on the rim. Father Carlo quietly swaps his for mine.

While we take a break to have our cake, I mention that I watched Pope Francis zipping around after Easter Mass on an electric scooter, shaking people's hands, utterly exposed to the crowd.

"Yes, and it drives the security people crazy." Father Carlo shares a story about the time Pope Francis needed to fill a new glasses prescription. Without telling security, he paid a visit to the optical shop. The optician was delighted but ultimately a bit disappointed. "Francis said to him, 'I want only the lens, no frames. Because I already have the frames.' He really said, 'I have the frames. I just need the lens.' " Father Carlo shakes his head, amused by the memory. "Can you believe?" His smile reveals a gap between his two front teeth. Because of the plain clothing and the regulation haircut — the absence of any indicators of personal style — the gap has the effect of an accidental intimacy, like bitten fingernails, or a bra strap that has slipped into view.

Presently Sandro returns with his tray to collect the plates and glasses. Father Carlo watches him retreat, then turns back to his guest. "Animals act according to their instinct, as you say about the coyote." He pronounces it *co-dee-oh-tay*, a most lyrical word. "Human beings, on the other side, have free will. They are responsible for the stewardship of creation. Their role is to help nature. Because we can study the system, and the animals cannot." He cites an example of reintroducing wolves into regions of Italy where deer and wild pigs were overpopulated, rather than culling the pigs and deer. "They asked the wolves to balance the ecosystem." He smiles. The gap again! I live for it.

I share the story of the Indian mongooses that were brought to Hawaii to kill rats in the sugarcane fields. What someone overlooked is that the rats are nocturnal, and the mongoose is diurnal. The mongooses ate a few rats and lots and lots of sea turtle eggs.

"Yes. Well." Father Carlo reaches for his briefcase. He has a train to catch. "In the complex systems of the world, we cannot predict the effect of all our actions. So? We have to act according to the principal of prudence."

Amen to that. Regarding the Italian wolves, I read later that they did indeed help lower the population of deer and wild pigs. They were fruitful and multiplied and began turning to ranchers' livestock for food. And then the ranchers began agitating for a cull. It is ever thus. In the words of National Wildlife Research Center public affairs specialist Gail Keirn, "When it comes to wildlife issues, seems like we've created a lot of our own problems."

Perhaps no place better knows the challenges and grand *doh!* moments of animal stewardship than my next stop, the lovely island nation of New Zealand.

(No Model.)

J. A. WILLIAMS.
ANIMAL TRAP.

No. 269,766.   Patented Dec. 26, 1882.

Witnesses: Inventor;
J. E. Clark.   Jas. A. Williams
W. H. Kerr   per
                                              F. A. Lehmann
                                                 attorney.

14
KILLING WITH KINDNESS: WHO CARES ABOUT A PEST?

To live in a penguin colony is to know no modesty. Anything you do — mate, preen, throw up fish for your young to eat — you do in plain view of the neighbors. The yellow-eyed penguin will have none of this. It builds a nest in the coastal tussock, out of sight from other yellow-eyed pairs. As with human suburbanites, the quest for space and privacy brings with it a longer commute. Every evening, yellow-eyed penguins of New Zealand's Otago Peninsula return from their labors at sea, crossing the beach, picking their way through the scrub, and trekking partway up a steep bluff to get home.

You can watch yellow-eyed penguin rush hour from a hide in a private reserve held by Elm Wildlife Tours. Shaun Templeton, Elm's dry-witted operations manager, leads today's outing. Shaun is youngish, with a shaved and well-tanned head. He has large brown eyes that remind me of a seal's, though maybe only in the context of this afternoon. Some beaches here have pinnipeds the way Coney Island once had humans.

Just past 5:00 p.m. the first penguin appears in the backwash of broken waves, bodysurfing. It lets the water carry it as far as possible and then, beached, it stands and crosses the strand in the deliberate plod that is a yellow-eyed penguin in a hurry. At the base of the bluff that rings the cove, it begins a series of considered uphill jumps, bending forward, pausing to cock the knees, then springing upward, its whole being applied to the task of moving one inch higher up the slope.

This is why, or partly why, the yellow-eyed is the world's most endangered penguin. The isolated nests and the long, exposed trek to reach them provide shopping and dining opportunities for predators: seals and sea lions, as always, but also newcomers — stoats (known in the United States as short-tailed weasels), rats, and feral cats. Only around four thousand yellow-eyed penguins remain in the world, forty-three of them currently at this bay. Stoats are known to have killed three chicks here already this season. Shaun keeps track, because he's here most weeks and because Elm contributes to the conservation efforts of the Yellow-eyed Penguin Trust.

It's dusk, the commute mostly over. We're watching the fur seals now.

"Have a look." Shaun indicates a loose scatter of bones: heads and spines of two barracuda, and some kind of octopus scaffolding. "A most impressive sea lion vomit." Sea lions eat their prey bones-and-all and then regurgitate what's undigestible, as owls do but less tidily. We happily note the absence of penguin solids.

Shaun makes his living from tourism but is by avocation closer to a naturalist. I don't know the precise parameters of that calling, but I think that if you find yourself applying the word *impressive* to any variety of animal vomit, you may be one. It has been a disastrous two decades for the yellow-eyed penguin. Shaun explains: On top of losing habitat to humans and getting tangled in their fishing gear, the birds have lately been killed off by disease — avian malaria and avian diphtheria — and starvation. As the sea has warmed, the bottom-dwelling fish that these birds eat have begun to move farther out into colder water that is also deeper. Yellow-eyed penguins can dive deep, but not as deep as some of these fish are now living. Perhaps more than anything else, though, it's the unnatural predators: stoat and company.

If the present rate of decline continues, the yellow-eyed penguin will likely be gone from the planet in ten or twenty years. It is difficult to be here watching them and not feel somewhat slammed by this information. What a thing to lose! Go look them up. The candy red beak, the pink go-go boots, the yellow mask angling back from the eyes. They're the Flash, they're 1970s Bowie! I don't mean to imply that adorable, showy species are of more value or somehow deserving of more concern. It's just . . . *damn.*

Up on the bluff, a pair is greeting each other. You can tell by the ruckus. The Maori name for the yellow-eyed penguin, *hoiho,* means "noise shouter." *Shhhhh,* you want to say. *The stoats will hear you! They'll wait until you leave and come for your children.*

It is not just the yellow-eyed penguin that is in trouble here. It's any flightless New Zealand species (and many that still have wings). For tens of millions of years, these islands had no native land predators, so birds who arrived here no longer had need for swift escape. Evolution gradually discontinued some species' wings, applying the energy to something more contributive to survival. Then the predators showed up — stowaways and introductions from other land masses. Each year, invasive species kill around 25 million native New Zealand birds: kiwis, most famously, but also kakapos, blue ducks, blue penguins, and keas (the carrion-eating mountain parrots we met in chapter 11). Stoats are efficient killers and ready tree climbers. They like eggs very much, but they will also kill and eat young birds. Each year 40 percent of all North Island brown kiwi chicks, for instance, are killed by stoats.

Stoats. Who invited them?

It began with rabbits. In 1863, homesick European settlers formed the Otago Acclimitisation Society, one of several such groups in New Zealand at the time, and released six rabbits into the Otago countryside. It was hoped, they explained, that "sportsmen and naturalists would be able to enjoy the activities that made the remembrance of their former home so dear."

What followed is nicely if perhaps hyperbolically summed up by one landholder's "rabbit arithmetic": $2 \times 3 = 9{,}000{,}000$. Two rabbits in three years equals 9 million rabbits. By 1876 most of Otago was overrun. The rabbits had no natural land predators, and the mild climate lengthened the breeding season. Rabbits ate the sheep pastures bare. Flocks starved. More than a million acres of Otago landholds were abandoned.

By 1881, government officials were moved to take action. They passed a Rabbit Nuisance Act, hired rabbit inspectors and "rabbiters" to shoot and poison. Stoats and ferrets were shipped in from Europe; nearly eight thousand of these "enemies of the rabbit" were released into the New Zealand countryside and protected by law.

But rabbits were only one of the species that wound up on the stoats' plate. The fierce tubular hunters quickly set to the avifauna: eggs, chicks, small adults. The march to genocide began. As of 2019, 79 percent of New Zealand's land-dwelling vertebrate species were classed as either threatened with, or at risk of, extinction. That includes 78 bird species and 89 reptile species.

In 2012, the New Zealand government again stepped in. Where once it had imported and protected stoats, now it endeavors to be rid of them. Predator Free 2050 (PF2050) is a Department of Conservation (DOC) effort to protect native biodiversity by eradicating the three invasive predators that most threaten it: stoats, rats, and brush-tail possums. (The year 2050 is when the DOC aims to have it done.) Citizens have been brought to the cause by energetic public outreach. Wander through any national park visitor center and you will see the Predator Free 2050 exhibit, with brochures and worrisome statistics and the requisite taxidermied stoat, needly teeth bared, one paw proprietarily placed on some savaged bird or violated egg.

Entire towns have taken the bait. Neighbors and farmers set out DOC traps by the hundreds. Success stories and tips are shared in a monthly newsletter that closes with the sign-off "Happy Trapping!"

Today has been a tour not just of Otago's birds but of the various models of DOC traps used to protect them. Each predator has its own challenges. "Cats like fresh meat," says Shaun, "so you need to rebait a lot."

Most places on earth have some feral cats, but this is well beyond *some feral cats*. Because along with stoats and ferrets, cats galore were set loose in the countryside to kill rabbits. Were in fact farmed for the job. And when demand outstripped supply, Dunedin lads were conscripted to prowl the city, stealing housecats.

So now, alongside the leering stoats in visitor centers and museums, you'll often see a taxidermied feral cat posed beside its stomach and the contents thereof: tiny paws, and feathers and bones, preserved in acrylic like some horrid paperweight that I kind of want on my desk.

On the walk back up the hill to the van, we pass a newer addition to the DOC arsenal, a self-resetting Goodnature A24. The bait sits at the end of a tube, behind a thin, flexible metal bar. When the head pushes the bar, it triggers the firing of a piston propelled by a carbon dioxide canister. Maybe you remember the unusual murder weapon used by Javier Bardem's character in *No Country for Old Men.* It's that sort of mechanism.

The stoat is famously hard to trap. "They don't like to stick their heads in things," Shaun says. Why then make a trap that's baited at the end of a tube? Because that way the head — the brain, more specifically — is positioned for a lethal strike and an instantaneous passing.

New Zealand is committed to eliminating invasive predators, but it is also committed to doing it humanely. No one person has contributed more to this effort than Bruce Warburton, whom I'm driving up to Christchurch tomorrow to see. Warburton helps design more humane traps, drafts animal welfare standards for traps, and tests traps to see whether they meet those standards. He is certainly the only New Zealander to hold professional affiliations with both the National Pest Control Agencies and the National Animal Welfare Advisory Committee.

What happened, in the end, to the rabbits? Through persistence, resistance, and multiplication, they're doing just fine. They survived the stoats and cats and rabbiters, and even the rabbit hemorrhagic virus smuggled in from Australia. Shaun says he's seen more than ever this year. With plenty of rabbits to eat, the stoats are thriving, too. "Stoats are through the roof," he says over his shoulder, from the driver's seat of the van that is taking us away from this beautiful and heartbreaking place.

Samantha Brown is a young biologist with freckles across her nose and an ear-pleasing Kiwi accent and extensive knowledge of how to kill quickly. Sam works with Bruce Warburton at Landcare Research, one of New Zealand's Crown Research Institutes. Their branch, on the outskirts of Christchurch, focuses on biodiversity and sustainability. That means lots of appealing job prospects for conservation majors. Though sometimes less so when they hear about the trap testing.

While we wait for Warburton to arrive, Sam cues up a video shot at the test facility, a few miles down the road. No tests are scheduled this week, which is, for me, less a disappointment than a relief. I appreciate and respect the good and emotionally wringing work the team does, but I wasn't sure I wanted to watch it live. There are things I too don't like to stick my head inside.

The setup is simple. A video camera on a tripod faces an observation pen. A person waits in the dark with a stopwatch and feelings it is hard for me to guess at. Sam points to a readout below the video frame. "You can see the time elapsing down here." Once the trap is sprung, the observer clicks the stopwatch. Humaneness, in the context of a lethal trap, is a function of speed: speed of death, yes, but more critically, speed to unconsciousness — to feeling and knowing nothing.

"This is a test of a Northland Regional Council cat trap," says Sam. The SA2 Kat trap is used not only on feral cats but also brushtail possums. The possums, native to

Australia, were released in New Zealand in the nineteenth century to establish a fur trade. They have thrived in their new home and multiplied and spread, feeding on and destroying vast numbers of trees that support native birds. On top of the estimated 21,000 tons of leaves and shoots they consume nightly, the brushtail possum also enjoys an egg.

On the screen, the time is displayed in tenths of seconds, increments that speed by unreadably fast, unless of course it's you in the trap. Sam and I sit silently through a half minute of don't-go-in-there, and then the terrible in-there. A colleague of Sam's rushes into the frame. Some part of me hopes he's been compelled to set the animal free. Sam narrates. "Here Grant is trying to look 'round the side of the head and touch the side of the eye." He's checking for the palpebral* blink reflex. When the reflex disappears, the animal is unconscious and the stopwatch is stopped.

* *Palpebra* means eyelid. Medicine has its own word for everything. The back of the knee? The popliteus. Earwax? Cerumen. *Hallux* means "big toe," and *mental protuberance* refers to "a prominent cranial bulge caused by memorizing too much Latin," oh, sorry, "chin."

The trap being tested in the video employs a metal bar that, when sprung, clamps down on the neck. While it's possible to kill a mouse or even a rat by snapping the neck in this manner, a larger animal in a trap like this one is killed by strangulation: the bar clamping shut the carotid arteries, depriving the brain of blood and thus oxygen. Suffocation plays a supporting role, as the bar may also shut the windpipe. Suffocation achieves the same end point as strangulation, but it takes longer, because it is cutting off air intake instead of blood flow. Thus the blood keeps circulating, and it takes a while for the oxygen already in the blood to be depleted. With a well-designed trap of this variety, oblivion arrives in 40 to 50 seconds. (Most film directors, though not Tarantino, accelerate strangulation to five or ten seconds, because, really, who wants to see more than that?)

A faster, kinder death could be delivered through a blow to the head. As a humane means of ending life, few things rival a bullet to the brain; thus "properly placed gunshot" is included as an acceptable mode of euthanasia in the guidelines issued by the American Veterinary Medical Association

(AVMA).* The world's first humane kill trap featured, in fact, a gun to the head. U.S. Patent No. 269,766, granted to James Alexander Williams of Fredonia, Texas, in 1882, featured a revolver set upright in a frame. In the illustration accompanying the patent, the barrel is pointed at some variety of rodent emerging from a burrow. Varmint steps on rod, rod pushes trigger. You don't get the sense, in reading the patent details, that Mr. Williams cared much about safety or humaneness or even rodent control: "This invention may also be used in connection with a door or window, so as to kill any person or thing opening the door or window to which it is attached."

So why are the makers of the SA2 Kat trap not trying to hit the skull? Because a

---

* As is decapitation. Humaneness was in fact Monsieur Guillotine's motivation. Scientific supply catalogs used to list an item called the "small animal guillotine." These days, at least online, you can only buy them used, perhaps because they attracted negative attention. For regardless of how humane a guillotine may be, it entails cutting off a head, and that is hard to stomach. And now I'm going to say this: if you want to sell a used rodent guillotine on eBay, for god's sake clean the blade before you take the picture.

humane head strike needs to be carefully placed. Carotid occlusion affords more leeway with body size and position. Also, Sam is saying, for the bar to hit with force sufficient to deliver a lethal, humane blow, one starts running into (human) safety issues. It would take some good strength to pull back the bar to set the thing, and if it got away and snapped shut on someone's fingers, bones would break.

Humane traps for stoats are among those designed to impact the head. This is because these animals have unusually strong, muscular necks, and the thick muscles protect the arteries. Also, the strength those muscles supply can enable the stoat to pull free.

Sam clicks a link for a video of a test of a stoat trap called a "modified Victor." This is the classic wooden Victor brand snap trap for rats, upgraded (by Warburton) for humanely killing stoats. A molded plastic hood, or shroud, is screwed in place over the bait to guide the head in from the proper direction and to the proper depth. The bar strikes at the latitude of the ears, and unconsciousness, Sam is saying, is almost instant.

"You can see it's started to go limp already." My brain offers a more benign interpretation of what it's taking in. *Look, it's one of those fur boas ladies wore to be classy in the nineteen forties.*

When I was twenty-two, I lived in a mouse-infested apartment building. The Mouse Death Count — lettered in German Gothic and stuck on the refrigerator by my graphic designer roommate — was up to thirty-two hash marks by the time I moved out. The landlord didn't allow cats, so we bought traps, the classic, cheap wooden Victor snap trap, the setting and disposal of which were my chore. I set them without much thought, assuming that they kill instantly because they hit the head or the neck. I was always taken aback by a mouse struck elsewhere: the one who came in from the side and was pinned by the shoulder. Another that changed its mind and got pinched by the snout while backing away.

According to University of Guelph bioscientist Georgia Mason's thorough, unblinking comparison of the humaneness of various rodent control methods, this sort of thing happens 4 percent of the time with a Victor snap trap. That's actually quite good, certainly compared to a competitor that, according to Mason, caught mice by the legs or tail 57 percent of the time. Mason studies behavioral biology at the university's Campbell Centre for the Study of Animal

Welfare, and her comparison study ran in the journal *Animal Welfare.* She is Canada's Bruce Warburton. She's outstanding.

These days the Victor company sells its own version of Warburton's "modified Victor," a snap trap called the Quick-Kill.* I commend them for this, but am disappointed to see that they continue to sell glue traps in their 2019 product catalog.† On top of the protracted torment of being glued in place, the animal may, Mason writes, rip off pieces of skin or gnaw through its limbs trying to escape. A professional pest control person should be checking the traps daily and humanely killing any rodent that's been caught, but Victor and others sell glue traps online to anyone, and what homeowner is

---

* Not to be confused with Victor Clean-Kill, Fast-Kill, Smart-Kill, or Power-Kill traps. All of them trademarked. Victor legal staff have basically trademarked the entire universe of small rodent killing: Kill Bar, Kill Gate, Kill Vault, Kill-Point, Multi-Kill, and even, if their catalog copy is to be believed, Mulit-Kill are all Victor trademarks.

† A lushly photographed 44-pager called Rodent Free Living. A number of pest control companies have adopted this term, as though mice and rats were a lifestyle choice, or something you go to rehab for. *I've been rodent-free for six years now.*

going to tackle that? So millions of mice and rats are left, adhered, to die slowly of dehydration. Luckier, Mason says, are the ones who get their snout stuck to the tray in the initial struggle and suffocate.

Glue traps are illegal in New Zealand and parts of Europe. I emailed the Victor product manager, asking whether the company had plans to stop selling glue traps. You won't believe this, but she didn't answer.

Sam's office door swings open. It's Warburton with a box of traps. A couple decades of his career in there. It clanks as he sets it down to shake my hand. Warburton has a genial, wry manner. He doesn't work at making everyone like him, but I imagine everyone does. He's an interesting hybrid: animal ethicist and hunter. I ask him how he got involved in the business of humaneness. It began, he says, when New Zealand's Society for the Prevention of Cruelty to Animals took an interest in a new, humane possum trap that they hoped could replace a particularly unkind leghold trap — key words: *serrated jaws.* (Possums are still commercially trapped for their fur, which is spun into wool.)* "They took it to our

* Possum merino is a heavenly soft wool blend.

minister of forestry, and he said, 'Well, how good is this trap?' and he came to us." *Us* meaning Landcare.

I guess I was less interested in the how than the why. I rephrase my question.

"I mean, they are pests," Warburton says, "but they're sentient animals.* They have the capacity to suffer. We have a duty of care to think about it, to minimize their suffering. That's just been my philosophy."

The traps in Warburton's box are mostly mechanical: trigger and thwack. Where's the newfangled stuff I've been reading about?

When I first arrived in New Zealand, I bought a pair of wonderful green possum merino gloves. I imagined peaceful possum flocks being sheared like sheep. Then Warburton explained that possum pile is too short to shear and is typically sliped instead. Sliping involves some kind of postmortem chemical depilatory. I still wear the gloves, though with diminished happiness.

* By definition of New Zealand's Animal Welfare Act, "sentient" applies to any animal with a nervous system that can relay stimuli from sensors around the body to the brain, plus a brain advanced enough to translate those signals into perceived sensations. That means all vertebrates, plus octopus, squid, crab, and lobster, but not, I'm relieved to report, oysters.

Before I arrived in New Zealand I spoke to a Predator Free 2050 researcher about a humane carbon dioxide trap under development for stoats and rats. The animal enters a tunnel and crosses an infrared beam, triggering doors at either end to close and carbon dioxide to be expelled.

At the right concentration and flow rate, carbon dioxide is believed to kill humanely. The gas is one of the AVMA's acceptable means of euthanasia. When a wildlife control operator live-traps Canada geese, say, or a raccoon in someone's attic, a CO_2 chamber may be the next stop. The operator may not mention this to the homeowner, and the homeowner may not ask, preferring to believe that the word *humane* on the operator's website means that the animal will be driven to some sunny woodland and set free. (Which may be less humane than the gas. See "The Fuzzy Trespasser: Resources for Homeowners," page 423, for advice on this.)

The humaneness of carbon dioxide was a subject of renewed debate at the 2018 AVMA Humane Endings Symposium. This is an animal euthanasia research conference held regularly near Chicago in November. Appealing! Here is the dilemma. It is elevated carbon dioxide levels in the blood, not low oxygen levels, that cause an organism to breathe harder. To avoid causing breathlessness and panic, you want the end to come quickly. But when the concentration and flow rate are high enough to achieve that, the CO_2 may start to form an acid on contact with mucous membranes, and the animal may experience burning and choking sensations. Getting it right is tricky.

Another newcomer to the humane endings arena is the electronic trap. These are box traps with double-decker electrified floor plates. When a visitor steps inside, the top plate tilts into contact with the lower one, completing the circuit and introducing the current into the visitor. Warburton tested one such device designed for possums. It was, he says, a mixed success in terms of animal welfare. "It works okay when the electrode plates are clean, but once they got a bit dirty, it cooked the wrists of the animal. So that wasn't very good at all." Warburton shares what seems to me to be a national distaste for euphemism and double-talk. His words are not offensive, just direct. Delivered flat, and at low volume. I won't be needing any exclamation points for the next few pages.

Georgia Mason weighed in on a commercially available version called the Rat

Zapper. Two thousand volts for two minutes. (As this gets underway, the homeowner receives a text: "RODENT CAUGHT." ROMANTIC DINNER RUINED.) Electrocution kills by disrupting the normal movements of the heart and diaphragm muscles. It's a death by ventricular fibrillation and respiratory distress, both of which starve the brain of oxygen. So painful are the muscle constrictions thought to be that humane livestock electrocution requires the shock to the body be preceded — or at least accompanied — by passing a current through the brain to induce unconsciousness. Whether these traps also do this is not known.

Mason ranked a well-designed electronic trap alongside a good snap trap as most humane of all the lethal options for rodents, largely because both kill quickly. The cutoff in New Zealand for an allowably humane kill trap is three minutes to irreversible unconsciousness. Having watched test videos of two traps that came in under a minute, three minutes strikes me as an eternity. Warburton overhears me saying this to Sam.

"Three minutes is not too bad," he says, putting traps back in the box. "The poisons are the tough ones." Here is the ugly reality of keeping stoats and possums and rats from causing extinctions. Predator Free 2050 relies more on aerial drops of poison bait than they do on the happy trapping of motivated citizens. Given the numbers and remoteness of some of the creatures that prey on native birds, traps are not going to get them all. And if you don't get them all, they will quickly bounce back — the rats especially.

New Zealand's DOC has been looking into better poisons: poisons that kill humanely, poisons that kill the targeted invasive species and no other creatures, poisons that don't build up on the land and in the animals that live off the land. And Landcare dutifully tests them. "It's quite hard on the staff," Warburton says. Because it's hard on the testees. Because now you're talking about hours or days, not minutes or seconds.

The tests are run near the facility where the traps are tested, a short drive from Landcare's offices. Warburton is holding car keys. "You wanted to see it?"

If you do an image search for "L pill," Google will present you with many close-up photographs of low-dose aspirin tablets marked with an *L*. This made me laugh out loud, because the L pill I was interested in was the kind issued to World War II operatives facing torture and possible spillage of top-secret beans. The *L* standing for *lethal.* L-pills, as the name was styled, contained potassium cyanide. The OSS, forerunner of the CIA, chose cyanide because it's a fast, easily concealed way to end a person's life.

In her consideration of the various rodent poisons, Georgia Mason ranks cyanide as the most humane. It suppresses central nervous system activity, and it interferes with the blood's ability to ferry oxygen to the cells, causing a sort of chemical asphyxiation. Mason cites two New Zealand studies of ingested cyanide, one indicating that possums lost consciousness after a minute to a minute and a half, the other finding that it took around five minutes. In the span of time before oblivion, there were muscle spasms, which would be painful, and some convulsions. Because the convulsions occurred after the EEG indicated a loss of consciousness, the animal wouldn't have been aware of them.

However, an onlooker would. Appearances matter. When states execute prisoners via a "cocktail" of drugs, they usually opt to include something to paralyze the muscles. Are they wishing to paralyze the breathing muscles or are they more concerned with the clenching, grimacing, spasming, convulsing muscles? I ran this by longtime death row assistant federal public defender Robin Konrad, former director of research and special projects with the Death Penalty Information Center. State officials have given her both reasons, she said, but it was her belief that, yes, they were seeking to avoid unpleasant visuals and the likely outcry that could result.

That convulsions are visually disturbing is the operative principle behind an unusual frightening agent called Avitrol. As used by farmers, the chemical is applied to a tiny percentage of a bait scattered in the fields, maybe one in a hundred pieces. The lottery loser who eats it takes to the air, flapping and squawking and soon thereafter convulsing grandly and dropping dead. The idea is that the rest of the flock, witnessing the spectacle, would be spooked and flee the farmer's field.

In 1975, Ontario's minister of the environment declared that only pesticides shown to be humane could be applied in the province. Though Avitrol is used as a frightening agent, it is a poison and so was included in the testing. The convulsions, said the University of Ottawa report on the results, happened after the point at which an EEG indicated a lack of awareness similar to that caused by dissociative anesthesia. The team deemed Avitrol humane but warned that the scientific evidence would "never change the opinion of those observing the effects."

About that, they were correct. On You-Tube, there is, or was, a video of a pigeon on Avitrol. Commenters assumed the bird was suffering and posted messages of pity and outrage. (At least the women did. Men's posts were more along the lines of: "Where can I buy this stuff?" And: "Maybe they shouldn't shit on everything and this wouldn't happen to them.") It seems the theatrical death throes of Avitrol victims are more upsetting to people than to other birds. In studies comparing its effectiveness with common frightening devices, the chemical placed last.

USDA Wildlife Services deploys a cyanide device on behalf of ranchers who complain about coyotes. The original was developed in the 1930s, and was known as the Humane Coyote Getter. It's a buried cyanide ejector, triggered by tugging on a bait that protrudes from the ground. Assuming it is jaws that do the tugging, the poison ejects directly into the mouth, à la an OSS cyanide pill.

You kind of knew there was going to be

trouble. In a U.S. Fish and Wildlife Service study spanning 1940 to 1941, the Getter compared favorably to leghold traps in terms of humaneness — *what wouldn't?* — and the toll on nontarget wildlife. But the Getter got seven cows and twenty-four pet dogs over the course of that year, and so began its undoing. Between 2013 and 2016, the updated version, called the M-44, killed twenty-two pets and farm animals, and it has, over the years, injured a number of humans, including a man who mistook it for a geodetic survey marker he wanted to steal. At least one lawsuit is pending, and four states have banned M-44s. Sayonara, cyanide.

Warburton and I are walking along rows of enclosures. Most are empty. A few possums are asleep in hanging burlap sacks, which, he says, remind them of the mother's pouch. We're talking about a common and inexpensive rat poison, an anticoagulant.

In small doses, anticoagulants are used to prevent blood clots — in, say, patients laid up in bed after surgery. (When my brother, Rip, was prescribed warfarin, he texted, "I'm taking rat poison," and he was.) In larger doses, anticoagulants interfere with the kind of clotting that patches tiny breaches in the capillaries — the normal wear and tear of the circulatory system. Animals that consume anticoagulants bleed to death internally, and that, Warburton says, is an unpleasant way to die.

"Depending on where the bleeding is, it can be associated with a lot of pain," he adds. Moreover, it's a slow death, occurring over a period of one to three days in rodents, and up to a week in possums. In the United States, some classes of anticoagulants have been restricted for use by professional exterminators or for island eradications of rodents threatening native wildlife.

Predator Free 2050 doesn't use anticoagulants. New Zealand's invasive predators are served up aerial drops of 1080-infused bait. Given 1080's legacy from World War II and chapter 8, I was surprised to learn this. Warburton says the poison's effects — on possums, specifically — fall midpoint on the humaneness spectrum. "They get nauseous in the last few hours, but it's not too bad." (The Denver Wildlife Research Laboratory, back in the day, found 1080's effects to vary widely by species, from "progressive depression" to — in dogs and other extremely sensitive species — "most violent epileptiform convulsions.")

With differing degrees of nastiness, 1080 kills a broad range of mammals and birds.

PF2050 baits are dyed to make them unappealing to birds, and New Zealand has almost no native mammals — just two bat species, neither interested in bait pellets. A body count published in the *New Zealand Journal of Ecology* determined the number of native birds killed in 1080 drops to be "negligible" — at any rate, not high enough to outweigh the benefits the birds derive from having their predators offed en masse.

New Zealand may have few native mammals, but it has seven introduced deer species — hunting stock shipped over by acclimitization societies and still hunted today. And 1080 kills the bejesus out of them. "The people who run 1080 operations get death threats from hunters," Warburton says, maneuvering a round-about. "They say 1080 is cruel to the deer, and then they go and stick arrows in them." Warburton half smiles. "I can say that, because I'm a hunter." To appease the hunters, deer repellents were developed and are added to 1080 bait in some areas.

Another concern with 1080 is secondary poisoning — the killing or sickening of creatures that might feed on the carcasses of stoats, possums, and rats. Pet dogs are the obvious example, but keas — the mountain parrots we encountered in chapter 11

— also feed on carrion and have been killed as well. The kea is one of the native species Predator Free 2050 seeks to protect.

So now the DOC needs a kea repellent, too. "These guys in here are actually next week going to be getting some 1080 with kea repellent," says Sam, who is running the trial and has caught up with us. "We need to make sure the possums and rats will still eat the bait." Anyone would agree that adding a kea repellent to your poison bait is an excellent thing to do, unless perhaps they, like me, had just made the acquaintance of one of the taste-testers.

Brushtail possums are fluffy, and they don't have the naked pink tail or the long snout of the American opossum. Their eyes are more forward in their face, like human eyes, or kitten eyes or the eyes of pretty much anything you'd find incredibly adorable.

"I know." Sam acknowledges my pout. "You just hope it will reduce kea deaths."

We move to another set of enclosures. "So this guy," Warburton says, "has been given a form of vitamin D, called cholecalciferol, that's under consideration as a replacement for 1080." Possums are particularly sensitive to it, and birds are not. "But it's not a very nice toxin, you see. It calcifies things."

Soft tissues. The heart. "And it takes quite a long time, and they stop eating. We were talking yesterday about whether we should just say, 'No, we are not doing any more trials.'"

Warburton and I say goodbye to Sam and head to the parking lot to drive back to the Landcare office building. Given what I've just seen and learned, the extent of public buy-in for Predator Free 2050 surprises me. A large 1080 operation may blanket 80,000 hectares (close to 200,000 acres). A government brochure illustrates the dispersal regime, using an image of a tennis court with five evenly spaced baits, as though someone had it in for Federer. At about five possums per hectare, that's 400,000 dead possums. And who knows how many dead stoats and rats. Plus all the deer and the occasional endangered bird. "Dead forest" is a term I've seen used — not by activists but by someone with the USDA — to describe the aftermath of a 1080 aerial campaign.

In light of New Zealand's widely touted commitment to the environment, I would have expected more widespread resistance.

"No one blinks an eye, because it's in the forest and it's at night," Warburton says. "If it was during the day and it was on pastureland, like this" — he tilts his head toward the car window — "we wouldn't be allowed to do it. I think it allows us to get a bit more social license than we deserve. That, and the fact that these animals are invasive. There's a pervasive media presence telling us that these things are pests and they're eating our forests, killing our birds, and there's just a general acceptance."

It's not just the media. Anti-predator propaganda turns up everywhere. Gift shops in New Zealand national parks sell jokey roadkill-shaped Squashed Possum chocolates. A popular children's book pits a scrappy band of endangered birds against a Grinch-faced stoat. ("There's no way any of us could fight him! We're doomed!")

Warburton rejects the stoat hate. "Stoats are amazing wee animals. They're incredible climbers, and they're incredible predators. They'll take on an animal bigger than themselves." Especially irksome to Warburton is the double standard applied to pets that kill endangered birds. Mostly cats. "I detest cats."* Says the winner of the Royal Society of New Zealand Bronze Medal for

---

* New Zealand says it like it is. On the brink of a cliff above the surf at Pancake Rocks is a sign warning tourists not to climb over the fence. It closes with "Don't be an idiot."

significant contributions to animal welfare research. Warburton would like to see the keeping of pet cats made illegal.

"And good luck with that, Bruce."

"I mean, you keep the one you have, but you don't get to replace it." Otherwise Predator Free 2050 is actually Predator Free Except for the Housecats That Decimate Endangered Bird Populations and the Dogs That Kill Adult Kiwis Unless You Give Them "Kiwi-Aversion Training" 2050. Which seems unfair to the possums and stoats. I keep recalling my husband, Ed, imagining what the possums say to each other. *Why do they hate us? Why? We give them the lovely wool for their gloves . . .*

Are there New Zealanders who advocate doing nothing?

"There are," Warburton says. "There are people who argue that if you leave it long enough, things will find a new balance. You'll lose some species, but others will adapt. Other people say we can manage [the predators] at certain sites." Meaning an island or a fenced area or something like the Otago Peninsula, with a lot of endangered wildlife (and wildlife tourism) to protect. "And then there's a suite of people who think that we can eradicate from the whole island."

Warburton stands somewhere in the middle. "From a practical standpoint, we can't afford to do the whole island. It's five hundred to a thousand New Zealand dollars per hectare. You can't roll that out over 26 million hectares. And the rats may come back." As they always do, from ships in port.

It strikes me that the PF2050 movement shares some DNA with that of the old acclimitization societies. There's some of that same desire to have the land around you look the way you have always known it to look, a belief in an ideal, static ecosystem. But ecosystems are always evolving. "There are botanists," Warburton says, "who don't like the deer because they eat the understory and modify the forest. But we used to have moas that ate the understory." Moas were similar to emus, but larger, and long ago hunted to extinction. "So they're trying to restore a forest that's sort of post-moa extinction and pre-deer."

In parts of New Zealand's South Island you run across signs warning of the dangers of "wilding conifers." Pine trees! A threat to lands and lifestyles! Using up scarce water! Altering iconic landscapes! From what I could tell, it's something of a lost cause. They were planted as windbreaks, and now they're all over. And quite lovely! Forgive me, National Wilding Conifer Control Programme. It's just hard to know where to draw lines. What to save, and at what cost. Out on the beach yesterday, I was ready to support whatever it took to prevent the extinction of yellow-eyed penguins. Today I'm less sure. It's hard to feel peaceful about the killing of some species in order to preserve others.

In part, it's the way it's done. Poison seems so 1945. Isn't there, by now, something better?

15
THE DISAPPEARING MOUSE: THE SCARY MAGIC OF GENE DRIVES

Everyone wants to eat a mouse. Hawks want to, coyotes want to, skunks, foxes, rats. A mouse is a nutritious morsel with no anatomical defenses: no venom or noxious exudations, no spines, no shell. A mouse's best hope is to get itself someplace safe, and do it fast. At that, the mouse excels. It can squeeze through a hole no bigger than its head. A motivated mouse can jump straight up to a height four times its body length. If I were a mouse, I could leap a wall twenty feet high with no running start. I could pass through the opening of my own mailbox.

I've seen the studies* and the video foot-

* Will Pitt, of the Smithsonian Conservation Biology Institute, documented it when he was with the NWRC. The house mice tested were all able to access food by passing through 13-millimeter holes — the average width of the local house mouse head. Less formal proof takes the form of

age. It's on the computer of Aaron Shiels, a wildlife biologist who has been working to create an escape-proof habitat for mice at the National Wildlife Research Center headquarters in Fort Collins, Colorado. The main feature is a room-sized simulated natural environment, or SNE ("the snee"), which Aaron is showing me this morning. Each mouse is implanted with an ID chip, and beneath the SNE's raised floor is a chip reader, ensuring every rodent is accounted for. The walls are smooth (unclimbable) plastic plates, screwed in place from the outside to keep mice from springing up and securing a clawhold on the screwheads, like rock climbers holding on by their fingertips to barely discernible protrusions in the rock. Metal flashing covers the seams of the SNE and the caulking in the corners of the outer

---

my friend Steph, who, one hot day, upon returning from a walk with her dog, grabbed a water bottle from the floor of her truck. She took a swig and discovered, in order, a fetid taste and a dead mouse inside the bottle. "I don't think you'll need a doctor," an advice nurse said upon hearing that she'd spat the water out. "But you may need a psychiatrist." For me, Steph dutifully went back and measured the head. It was exactly the size of the water bottle's opening.

room that contains it. Because a mouse can gnaw a hole anywhere it can wedge its teeth — in wood, plastic, cinder block, aluminum. The front teeth of a mouse are self-sharpening chisels. The back sides are softer than the fronts, so each time a mouse brings its teeth together, the hard enamel on the front of its lower incisors wears away the softer backs of the uppers, honing the edges. *Rodent* comes from the Latin for "gnaw." They do it well and fast, so well and so fast that they need to suck the insides of their cheeks through a gap in their teeth to close off their windpipe and keep the sawdust out.

The current residents of the SNE are mice of no great value or concern. The security features are being designed for a group of future occupants: mice genetically modified such that they produce only male offspring. And further engineered, through a process called a gene drive, to spread this trait far faster than it would spread naturally. Gene drives are one possible future of invasive species control, a potential alternative to strewing poison all over an island.

Like any genetic modification, gene drives make some people uncomfortable — not just among the public but among the scientific community. Most prominently, Jane Goodall. And so, before the engineering of the mice must come the engineering of the habitat that will keep them in. Part of Aaron's work has been to create such a place and to demonstrate to a high level of certainty that escape is impossible. Even for a mouse.

"Otherwise, I'm all over the news." In 2017, an elk with a highly contagious bacterial infection escaped from a USDA Veterinary Services facility next door to the NWRC. "People thought it was ours." Aaron has hazel eyes and shoulder-length red-brown hair that is, today, gathered in an elastic band. I just realized it was Aaron I saw earlier at the front gate. He had sped past on foot, ponytail aloft, while the guard checked my ID. He looked like a man crashing the gate, but now I know he's just a man who runs to work in the morning.

Of all the invasive species harassing and killing endangered island natives, the house mouse is not high on the fret list. They can be a problem for seabirds on islands that evolved without seabird predators, for instance the endlessly beleaguered albatrosses of Midway Atoll. (Wildlife cameras in 2015 began capturing ghastly scenes of mice feeding on albatrosses as they sat incubating their eggs.)

But seabird conservation is not the reason

mice were chosen as the gene-drive test species. Mice were chosen because science knows mice. You can't fiddle with an animal's genome without first decoding it. Additionally, mice can crank out a litter in a few weeks. Gene drives take effect over generations, so researchers who wish to have data before they reach retirement age prefer speedy breeders.

To date, only one mouse has escaped the SNE. It went down like a movie prison break. The mouse had burrowed deep into the sawdust shavings provided as bedding, and when staff came in to change out the shavings, the mouse was scooped up in the dustpan along with the bedding. He escaped in the laundry cart! (And was caught the next day in a trap outside the enclosure.)

The SNE is quiet at the moment. The mice are sleeping. They're part of a trial of the Goodnature A24, to see whether it also works, and works humanely, on mice. So far the challenge seems to be the bait. In a lush tropical environment, a bait has to be extremely tempting* to compete with the

* To eradicate invasive, bird-decimating brown tree snakes on Guam, the U.S. government has used a bait of dead neonatal mice (or "pinkies," in the pet snake community) laced with acetamino-

natural food sources. Aaron hands me a bottle of Goodnature bait, a chocolatey-coconutty goo that smells, but does not taste, delicious. It's like eating suntan lotion, I tell Aaron.

---

phen (or Tylenol, in the pain-relief community). But pinkies aren't cheap, and within three days they start to become unpalatable, even to a snake ("green coloration . . . followed by swelling, seepage, odor, and eventual rupture"). So began a fifteen-year search for a replacement bait. Because snakes don't chew and savor their food as we do, the thinking was that you could coat a cheap core material — sea sponge was tried, and gum rubber — with an irresistible attractant. A dozen or more wrappings and coatings failed to seduce captive snakes at Monell Chemical Senses Center in Philadelphia: Roquefort cheese, termite attractant, poultry fat, fetal pig skin, evening primrose oil, baby formula. (The researcher, Bruce Kimball, had expected resistance from the baby formula company, but they were all in. The Tylenol people have maintained a cool distance.) Eventually a winner emerged: "mouse butter" on a core of potted meat product. "We reverse-engineered Spam," Kimball told me with understandable pride. Spam is cheap, keeps for at least a week, and doesn't attract ants — and the snakes, like the humans of Guam, are inexplicably fond of it.

"You *tried* it?" His look combines horror, confusion, and pity. "You want some gum?"

Like many people, I have some qualms about genetic engineering and its possible future. Also like many people, I know diddly about how it works. My plan for this afternoon is to become someone who knows a little more than diddly. I'm scheduled to meet with the Center's wildlife genetics

---

The other challenge was to get the baits where the tree snakes were (and other creatures weren't). The pinkies had been strung with plastic parachutes, like toy army men, and dropped from helicopters into the canopy, where the parachute cords would catch on branches. Your taxpayer dollars making my day. However, hand-attaching six thread-thin parachute cords to each bait proved "tedious," and to cover all of Guam, you'd need two million baits. Now they use an "aerial broadcast system," a sort of helicopter-mounted machine gun that shoots potted-meat baits with biodegradable cornstarch streamers to entangle them in the trees.

The upshot? While brown tree snakes have quickly eradicated nine out of twelve native Guam bird species, the combined might of the U.S. Navy, the USDA, and U.S. Department of Fish and Wildlife has thus far failed to return the favor.

staff, upstairs in the Long Speak Room, which is an amusingly apt name for a government conference room (except that it isn't — a realization that will dawn when I take note of the plaque by the door, which reads: Longs Peak Room).

I'm in the main lobby now, waiting for my escort. There is, yes of course there is, taxidermy. A family of monk parakeets is posed in and around a nest in the top segment of a power pole. The diorama takes up most of the surface of the small table beside my chair, forcing me to place my coffee directly beneath the birds and fostering a vague sense of unease.

My escort arrives and we make our way toward the stairwell. The corridors are hung with mounted research posters and color blowups of photogenic "nuisance" species: cormorants, ground squirrels, beavers. Wildlife management agencies and pest control websites do this too, and it always hits me as slightly off — it would be like the FBI decorating the hallways with headshots of good-looking federal criminals.

Up in the Longs Peak Room, I take a seat next to Toni Piaggio, a conservation genetics specialist. Her own personal genetics have provided well. She received: graceful cheekbones, blinding intelligence, glossy black curls, deep reserves of patience. Toni introduces a young colleague, Kevin Oh, also a geneticist.

Before we get to the *should* and *could* of gene drives, a wobbly stab at the *how*. A gene drive is two unique manipulations. First, there is the one people are familiar with, at least in a general way: genetic modification, the GM in GMO. This is done using a technology called CRISPR-Cas (or CRISPER, for short). A target gene is selected — say, a gene for a trait that enables a mosquito to carry malaria — and then what Kevin calls the "molecular scissors" of CRISPR cut out and/or replace the target gene. The edit is done, in this case, early in the embryo's growth — when it's just a few dozen cells — so that the modified genome will be copied into every new cell going forward.

CRISPR-Cas is a natural element of bacteria, a part of the mechanism that serves as their defense against viruses called phages. This defense system includes an enzyme that cuts up the virus's DNA, and as it does so, it retains a memory of it — a "molecular bar code," as Kevin puts it. So if the virus returns, its specific genetic sequence will be recognized and cut up. Geneticists have harnessed the CRISPR scan-and-snip system as a way of precision-targeting and editing DNA.

"But how does the enzyme get *in* there?" I'm whining.

"They're literally injecting mouse embryos," says Kevin.

"Using, like, super-tiny dollhouse hypodermics?" I want to see these.

Toni steps in to move things along. "There are different methods. We sort of flood the embryo in its petri dish."

The enzyme gets in. It scans, it finds a match, and *beep,* it does its thing.

So let's say that you've managed, with your scissors and your bar code reader and your dollhouse drug paraphernalia, to manipulate the genome of a group of mice. And that now these mice cannot produce female offspring. If you set enough of them free on an invasive mouse-plagued island, the population will start to dwindle.

"Enough" is the challenge. This is where the gene drive element comes in. With normal Mendelian inheritance, this new trait would show up in 50 percent of the offspring, because half the offspring's genetics are contributed by the male. What a gene drive aims to do is deliver the machinery to make that gene 100 percent heritable. So that now all mice born to the gene-drive mouse will carry the trait. A successful gene drive would speed the time it takes for a trait to spread through a population.

Here's a hitch. In order to swamp an island population to a point where the gene-drive animals make sufficient inroads, scientists will have to release large numbers of the very species they seek to be rid of. Depending on the size of the island's invasive population, it may take a considerable number of lab-born gene-drive animals to tip the balance. So before a gene drive can improve the situation, it will temporarily make it worse. Thus the first step, with rodents, would likely be an aerial drop of a rodenticide — the practice we're trying to get away from. The gene-drive rodents would be released afterward as a sort of mop-up and maintenance program. So that any survivors, and any newly introduced future rodents, will peter out — rather than building up their numbers again and necessitating another hail of poison.

So far, the gene-drive element is proving tricky. The gene-drive mechanism isn't always copying correctly. And there seems to be a narrow window within the embryo's development during which the manipulation will take. Too early and it kills it; too late and it doesn't transform. The sexual practices of wild mice may also pose a challenge. Polyandry — whereby one litter may include offspring fertilized by multiple males — was recently shown to be more common among wild mice than previously thought. Thus it may take longer and require more gene-drive males to overtake the population.

On a more basic level, there's a possibility that the invasive residents won't breed with the genetically modified newcomers. In nature, mice from different islands or regions of the world begin to evolve separately. Not necessarily to the extent that they're different subspecies, but different enough that they may not breed with one another. "You create these mice in the lab and then you have to make sure the wild mice on the island find them sexy," says Katherine Horak, a toxicologist sitting a few seats to my right. Laboratory *Mus musculus* are surprisingly different from wild *Mus musculus*. "Lab mice want to just sit there and hang out in your hand. The first time I worked with a wild mouse," Horak recalls, "I was like, *What is that?* It wanted to jump up and bite my face off." (Thus the need for a high-security habitat.) One of the first things that will happen in the SNE are mating trials — back-crossing the gene-drive lab mice with wild mice to create a strain that's sufficiently sexy to the invasives one wishes to control.

The big concern with gene-drive organisms is that they'll make their way beyond the area — and the population — they're intended to control. And that the locals wherever they end up will *not* hesitate to breed with them. Say you create a gene-drive "daughterless" feral pig and one of them mates with a domestic pig. Pig farmers would be undelighted. This is one reason scientists would plan to start with a physically isolated population — invasive rodents on a remote, uninhabited island, for instance. (Preferably an island no ships visit, because mice and rats are notorious stowaways.)

There is a way to avoid the concerning scenario. The same genetic drift that could prevent mice from different land masses from mating can be harnessed as a safety feature. Geneticists can target a genetic bar code unique to the population of the particular island or region. "So CRISPR can sit down and cut at this place in the genome that we only find in this population of mice," Toni says. "So even if someone nefariously transported those mice somewhere" — or they transported themselves as stowaways — "we wouldn't have to worry about it migrating into the local population." In other reassuring news, research out of the University of California, San Diego, suggests it may be possible to halt or even reverse a gene drive. In a paper published in the fall of 2020, two new gene-drive control mechanisms were shown to work in fruit flies.

Katherine Horak has been working on something altogether different and less anxiety-provoking than gene drives. It's called interfering RNA, or RNAi. Riddle me this: It's a bait that kills, yet it contains no poison. It's a species-specific genetic solution, yet it doesn't modify the target species' genome. That makes it an appealing package: doesn't effect nontarget animals, safe for the environment, can't go rogue. RNAi is based on a mechanism all organisms have: enzymes that patrol for viral RNA and destroy it. So you would choose a protein critical to your target animal's life processes, dress it up like viral RNA, and let the interference mechanism destroy it. There are of course hundreds of proteins critical to life. Horak will look for one that would end it swiftly and without suffering, probably something neurological or cardiac.

The challenge with RNAi bait is that

you're sending delicate strings of genetic code through the acids and enzymes of a digestive tract. Horak is working with biochemists to design a carrier molecule. That will take a while, as will getting RNAi registered with the EPA. It's at least a decade away from being something you'll see at Home Depot.

The novelty of the approach may trip it up. "There's a lot of talk in my world about perceived risk and actual risk," Horak says. "People are more comfortable with the actual risk of anticoagulant rodenticides, which will kill anything if they eat enough. But because we've done it for so many years, that level of risk is somehow acceptable." (In some places, and for some people, anyway.) "But the risks of RNAi are new risks, and so there's hesitation around those risks."

RNAi will face the inevitable challenge of any island eradication effort that relies on bait. The holdouts. The rodents that never encountered any bait. (Or, with poison bait, the ones that nibbled it, ate enough to feel sick but not die, and from then on steer clear.) Agencies can end up spending as much money tracking down the last ten of an invasive species — and monitoring for eleven, twelve, thirteen — than they spent eradicating the first ten thousand. This is going on right now in California's Sacramento–San Joaquin River Delta, where nutria have been proliferating. Nutria are similar to beavers — big swimming rodents that like to manipulate the landscape in ways that can make them unpopular. But nutria breed faster, and they're invasive. To find the holdouts, the California Department of Fish and Wildlife has been releasing neutered "Judas nutrias" — radio-collared individuals turned loose to betray their hidden kin.

A gene drive would have the rodents eradicate themselves. Without death or pain, and without killing any nontarget species.

And yet.

Here are some species the EPA, the USDA, and the Department of Health and Human Services consider "pests": chipmunks, bears, raccoons, foxes, coyotes, skunks, flying squirrels, tree squirrels, little brown bats, rattlesnakes, coral snakes, cliff swallows, crows, house finches, turkey vultures, black vultures, and mute swans.

This is what troubles me, I say to Aaron. We're back in his building now, watching rows of mice in stacked Plexiglas habitats, a Hollywood Squares of mice. Paul Lynde is doing backflips off one wall. What if a government agency eventually decides to go forward with gene drives on these other "pests"? What if economic considerations start to determine which species are next? What then? So long, pocket gophers? Toodle-oo, "nuisance beaver"? Right now, the focus is island conservation. It's a more appealing and less worrisome application: saving endangered species in a geographically isolated location. So you try it out there, and it works well, and the native species recover and there's good press. Now what? Where does the line get drawn, and who draws it? Let's remember: The National Wildlife Research Center is part of the USDA. It's not a conservation organization. "Aaron, the end point here, the ultimate target, is agricultural pests? Right?"

"That has been part of the discussion," he allows.

This is where it gets scary to me. We've seen what happens when the deciding factor is agriculture's bottom line. Will gene drive be a tidier rendition of the poisoning, shooting, trapping, bombing, wipe-'em-out campaigns of past centuries?

Aaron agrees that the decisions can't be just financial. "It has to be tied to ethics. We feel like we've taken a lot of steps to be sure this is accepted on a lot of fronts and we're not trying to go to third world countries to test it out." But the United States is surely not the only country working on gene drive in mammals. If we're on it, China is too. And China has not demonstrated a comforting abundance of oversight in the realm of genetic engineering.

Aaron was at the GBIRd meeting where Jane Goodall called for a moratorium on gene-drive research. GBIRd stands for Genetic Biocontrol of Invasive Rodents; it's a consortium of five U.S. and Australian government agencies and universities, plus the nonprofit Island Conservation. I ask him what Goodall's stated objection was. (Efforts to communicate directly with her were unsuccessful.)

"I think the concern is that the technology and the ability of people to experiment with it is moving way too fast, and the only way you're going to slow it down is to shut it down completely," he says. "And I think it's good. If someone reputable like Jane takes a stand, people will stop and think, *Maybe we need to establish some guidelines that we'll all follow.*"

Yes, please. Guidelines. Imagine if, rather than reducing the population or geographic distribution of a species, a gene drive wiped a species off a whole continent. Or an ecosystem was changed in some unanticipated and catastrophic way. It's the unknown unknowns that trouble some biologists. I spoke with Will Pitt, a former project leader at NWRC and now deputy director of the Smithsonian Conservation Biology Institute. Rather than any specific fallout scenario, he expressed a general wariness. "People always say, 'We've thought of everything that might be a problem.' Well, it's probably one of the things you haven't thought of that's going to be a problem."

In a corner habitat on the top row, Charles Nelson Reilly is pirouetting on his hind legs. *Look how clever we are. See how we dance! Don't wipe us out!* Personally I would hate for mice to disappear. As would species for whom they're a common bill of fare. Who knows what defenseless small meat they'd turn to instead? But I suspect I'm not in the majority opinion as regards small rodents. I think plenty of people would be a-okay with, say, a global *Mus musculus* extinction.

"Right?"

"You mean if you asked a farmer or rancher who has a mouse problem?" Aaron chews his gum and considers this. "And you said, 'What if mice were eliminated from the planet, even from places where they have some important function?' "

"Yeah."

"Yeah, they might be like, 'I don't care.' "

Aaron knows a Big Agriculture guy with a lot of mice on his property. His name is Roger and he runs a feed lot where beef and dairy operations send their cattle to be raised. They're fed different diets depending on what they're being raised for, milk or beef or breeding more beef. The mice enjoy all of the diets. When Aaron needs wild mice for the SNE, he drives out to Roger's place. Roger surely has opinions about bothersome rodents and what their fate should be. Aaron agrees to drive out there with me after lunch.

Roger arrives to greet us driving a bulldozer-sized forklift. His cowboy hat is white felt, and the rest is mostly denim. He steps down and extends a hand. His grip is strong, but not in the manner of a person who's been coached on the importance of a firm hand-shake. More in the manner of a person who uses hand tools a lot. "Glad to meet you," Roger says.

Aaron hasn't been out here in a while, so he reintroduces himself. "I know who you are," Roger says. "You guys are the ones that let that elk out." Aaron lets this drift.

We follow the hat into the interior of Roger's grain elevator. As our eyes adjust to the dim, we start to see them. Every half minute or so, a mouse races along the base of a wall or shoots across the floor and disappears beneath a pile of metal machine parts. They say if you see twenty mice, there are two hundred more you're not seeing.

We go back out into the sun to continue the conversation. Above us is a silo of cracked corn and, I'm guessing, those other two hundred mice.

"Nah, that's pretty much mouseproof," Roger says. The top button on his shirt is open, and a long white chest hair quivers when the wind rises. "They could go up there, but why? They don't have to go very far to find something to eat." He scuffs the ground with one boot. There's enough spilled corn that the driveway crunched like gravel when we pulled up.

Other feed ingredients are stored out in the open. At the end of the drive is a low mountain range of brewer's grain and barley hops. I ask Roger to estimate the percentage that's lost to mice.

"Well, it comes in twenty-five-ton lots. How do you know if mice ate fifty pounds of that?" He removes his hat with one hand

and with the other, wipes away sweat. His face is tanned up to where the hat begins, then not. "In the grand scheme of things, the wind probably blows away more than that. You know, so. I'm not sure that's a huge problem."

Roger's quibble with the mice is that they like to nest in the engines of his vehicles and sometimes they chew the wiring. But he doesn't set out traps or poison. "I try to keep barn cats. Though they keep going out there on the yellow line and getting run over. Or the barn owls get them."

I ask if he puts up nest boxes to encourage the barn owls, which also eat mice. This is a dumb question. Roger has barns. He doesn't need boxes. Though he has heard about the practice. "They're doing that out in California. Man, they eat a lot of mice." He surmises the reason he has no rat problem is that the foxes around his farm keep the population down. Probably so. In the late 1950s, overexuberant slaughter of foxes and coyotes in Oregon contributed to a massive mouse infestation. In California, circa 1918, one bounty program begat a second: three cents per ground squirrel tail or, in some counties, scalp.*

* Unscrupulous bounty collectors would cash in

412

In the sky over a corral of Holsteins, twenty or so black birds wheel east. Aaron asks about bird-hazing strategies. "There's guys that'll come in and shoot at the starlings," Roger says. He doesn't use them, he adds, because it's not effective. The birds take off, circle around, and quickly come back. "It's more of a psychological benefit. To feel like you're doing something." He watches the birds disappear behind a stand of trees. "It's not a huge problem."

I'd like to end this book right here in Roger's smelly, baking cattle feedlot. The man in the big white hat gives me hope. To me, he represents a possible future where people may be frustrated by wild animals that get up in their business but they're living with them. In that possible future, people's reaction to the damage brought about by wildlife is something akin to ac-

---

twice: scalps in one county and tails in another. Some would fashion faux tails by wrapping portions of the hide around a skinny stick. Still others would cut off the tails and set the rodent free to breed yet more tails. In considering a bounty program for invasive brown tree snakes in Guam, officials worried that people might be tempted to release them on islands where they did not yet exist, in order to create a novel income source.

ceptance. Or maybe resignation is closer. Anyway, something far short of the conscience-free rush to annihilation that characterized previous decades and centuries. If people are able to step outside the anger, they may find that more humane approaches are also more effective.

There are plenty of farmers and ranchers more progressive than Roger, and that's precisely why he makes me hopeful. He's big agriculture, not small organic, and yet he gets it. Without ever uttering the words, he's practicing coexistence and biocontrol. The feed he loses to rodents and birds is part of the cost of doing business. Perhaps the model should be shoplifting. Supermarkets and chain stores don't poison shoplifters; they come up with better ways to outsmart them.

Before we leave, Roger shows us around the feed lots. The breeding bulls are on a maintenance ration. Roger takes a handful from a trough and holds it out for me to smell. We walk on. "Across here are the commercial beef cattle." They fatten up on corn. "Real high-calorie, high-carb."

The fat cattle stand at the fence, swishing their tails and staring. *You're all on about the mice, but what about us?*

"They'll go to JDA or Cargill for slaughter," Roger adds casually. "Probably sixty days from now." Because people like me want to have their hamburgers. *Only once or twice a year,* I want to say. But I know that's a lame defense. It's not the quantity that matters, it's the statement you make or don't make. When you tell people you don't eat beef — or would never use a glue trap — you make the alternative a little less comfortable for them. You keep it from being a thing they give no thought to.

For centuries, people have killed trespassing wildlife — or brought in someone to do it for them — without compunction and with scant thought to whether it's done humanely. We have detailed protocols for the ethical treatment and humane "euthanizing" of laboratory rats and mice, but no formal standards exist for the rodents or raccoons in our homes and yards. We leave the details to the exterminators and the "wildlife control operators," the latter a profession that got rolling when the bottom dropped out of the fur market and trappers realized they could make better money getting squirrels out of people's attics.

Rodents are a good bellwether. If people can be less cruel to rats, if it even crosses their minds to be less cruel to a rat, then things are heading in a good direction. Not just good for rats, maybe also good for humans. "If man can be taught to respect the home of the worm," wrote the nineteenth-century historian Léon Ménabréa, "how much more ought he to regard that of his fellow man."

A few months after I returned from Colorado, I was reading a book outside. I happened to look up as a roof rat ran across the end of the deck. My immediate impulse was to drive to the hardware store and buy a snap trap. But I didn't. How could I? Little Miss Coexistence. Put your money where your yap is. Besides, as I knew, removing one raises a VACANCY sign for another.* My neighbor livetraps the squirrels that raid her peach tree and lets them go in a neighborhood park. She has been doing this, Sisyphus-like, the entire decade

* Or on a more magisterial scale, "I know from my own estate in Yorkshire. . . . as soon as one kills 300 or 400 grey squirrels in any given month a similar number come . . . and take their places." — the Earl of Feversham, parliamentary secretary of the Ministry of Agriculture and Fisheries, at a meeting of the House of Lords to discuss the Grey Squirrels (Prohibition of Importation and Keeping) Order of 1937.

we've lived next door to each other.

A few days later, heading down the stairs from the deck, I saw the rat again. He was running down a tree branch, with a loquat in his jaws. Our eyes met. He froze. I froze. He dropped his fruit. Seen head-on, the naked tail hidden from view, the roof rat is certifiably cute. The species is smaller than the Norway rat, with fur of a warmer, prettier brown. It's a squirrel without tail fluff. This guy was fully as adorable as the ground squirrels that run around the bayside park where I hike. (And if history is any indication, less likely to pass on disease.) I went on down the steps and put my load in the washer and forgot about the rat.

A week later, I heard something moving inside a wall. "Your little friend is going to chew through the wiring and set the house on fire," said Ed. I told him I wanted to figure out how it was getting in and practice "exclusion." He gave me a week to figure it out.

I set up my wildlife camera in different places around the outside of the house, and we figured out where the rat was getting in. Ed filled the gap, and that was that. The noises stopped. I continued to see the rat around, mostly on the camera, but once or twice our paths would cross. I'd nod hello, and we'd go about our days.

ACKNOWLEDGMENTS

While working on this project, I would occasionally encounter the term "vertebrate pest." I don't much care for it, because it reduces an animal to its role in the context of human enterprise. However, there is one mammal for whom the term seems fair and apt, and that is me. For kindness and forbearance in the face of ceaseless pestering, a few people must be singled out. Bowed down to. Carried high by the cheering crowd. They are: Stewart Breck, Justin Dellinger, Travis DeVault, André Frijters, Joel Kline, Dipanjan Naha, Aaron Shiels, Bruce Warburton, and Dazy Weymer. You made this book, as much as I did. And in return I have those two words that never really capture the feelings they represent. Thank you.

Though they are less visible on the page, the following people also welcomed me and shared their time and knowledge, often with little or no advance warning. Giant neon thanks to Samantha Brown, Father Carlo Casalone, Aaron Koss-Young, Charlie Martin, Dean McGeough, Nico Nijenhuis, Toni Piaggio, Qamar Qureshi, Saroj Raj, Tom Seamans, Shaun Templeton, Kurtis Tesch, Rafael Tornini, R. B. S Tyagi, and Tina White.

Kim Annis, Jonathan Clemente, Bradley Cohen, Sarah Courchesne, Doug Eckery, Julie Carol Ellis, Esteban Fernandez-Juricic, Dave Garshelis, Katherine Horak, John Humphrey, Bruce Kimball, Mario Klip, Page Klug, Tim Manley, Stella McMillin, Vicky Monroe, Julie Oakes, Seth Pincus, William Pitt, Samantha Pollak, Heather Reich, Virginia Roxas-Duncan, Shane Siers, Steve Smith, Peter Tira, Catherine Vande-Voort, Harry Wetherbee, Kate Wilmot, and Bonnie Yates: if I did not achieve full pest status, I was, at least for an hour or two, a thing buzzing in your ear. Thank you for not waving me away.

Tim Bibby, Paul Deckers, Carol Glatz, Thaddeus Jones, Gail Keirn, Kirsten Macintyre, Fabrizio Mastrofini, Heather Steere, Kevin Van Damme, and Brian Wakeling: your help made possible entire chapters that would otherwise not exist. My thanks to you. Keli Hendricks, John Griffin, John

Hadidian, and Kellie Nicholas, you put things in context both historical and political, and I deeply appreciate your insights. For allowing me to infest your inbox, thank you, John Anderson, Meera Bhatia, Johan Elmberg, Ann Filmer, Robin Konrad, Georgia Mason, Christina Meister, Sanath Muliya, and George Smith.

My reporting took me to places where I did not understand the language and culture. My gratitude to Raffaella Buschiazzo and Charles Lansdorp for help with translation and interpreting. Nilanjana Bhowmick, Aritra Naha, and Shweta Singh, to each of you: your skillful ear and quick mind added nuance to my reporting, and your companionship made me feel at home when home was far away.

To Jill Bialosky and Jay Mandel: twenty years and seven books down the line, here I am, thanking you again. Yet it never feels like enough. Because *everything always goes well.* How often, in publishing — in any business? in life! — does that happen? Contributing in other immeasurable ways to the ongoing W. W. Norton miracle are the following excellent humans: Steve Attardo, Louise Brockett, Steve Colca, Brendan Curry, Ingsu Liu, Erin Lovett, Meredith McGinnis, Stephanie Romeo, and Drew

Weitman.

I am indebted to Janet Byrne for her graceful and expert copy-editing and for the patience, tact, and zeal with which she checked my work. Few are as good as she (as *her?* Janet! Help!).

A nod to Carlton Engelhardt for the emus and to Andy Karam for the stilettos. Cynthia, thank you for introducing me to Nila. Jeff, thank you for listening and for thinking it was a good idea, and Jesse, thank you for the New Zealand connections and hospitality. Steph, thank you for adding monkey-time to the itinerary. And Ed, always Ed, thank you for everything.

THE FUZZY TRESPASSER RESOURCES FOR HOMEOWNERS

The Humane Society of the United States (HSUS) has a helpful "What to Do About" series on its website. Included are strategies for resolving — or, better, preventing — problems with urban and suburban wildlife species, among them, bats, bears, Canada geese, chipmunks, coyotes, crows, deer, foxes, mice, opossums, pigeons, rabbits, raccoons, rats, skunks, snakes, squirrels, sparrows, starlings, wild turkeys, and woodchucks. https://www.humanesociety.org/resource/wildlife-management-solutions

More good advice can be found in the "Living in Harmony with Wildlife" series on the PETA website. Included are bats, geese, mice, chipmunks, pigeons, raccoons, skunks, squirrels, rabbits, and rats. https://www.peta.org/issues/wildlife/living-harmony-wildlife/

If wild animals have begun raising a family

in an attic or crawl space, you will need professional help to undertake a humane eviction of both mother and young. Before you make any calls, I recommend reading the HSUS "Choosing a Wildlife Control Company" web page. Getting them out requires specialized know-how.

As does releasing them afterward. Best practice these days is called "on-site" release. Once the operator has helped you seal any entry points and eliminate or close off other appealing nesting sites, the animals are released inside their home range — that is, your property. Driving them to a nearby woods or park sounds humane but likely isn't. "The squirrels did not fare well," concluded a study by University of Maryland and HSUS researchers who radio-collared thirty-eight gray squirrels and relocated them inside nearby Patuxent Research Refuge. Seventeen wound up as carcasses, or a skull or piece of fluffy tail lying by a collar, or just collars — two "with tooth marks on them" and one inside a fox den. The remaining eighteen squirrels disappeared, on average within eleven days, fate unknown. Raccoons, by one study, fare better, but some states prohibit the practice, because rabies virus could be relocated at the same time.

Good news for homeowners and rodents: more and more pest control companies are offering "exclusion" as an alternative to poison bait boxes or trapping. It's a matter of locating all the (surprisingly small) crevices and gaps where mice, rats, or squirrels could gain entry and then filling them with something rustproof that the animals can't easily gnaw through: steel wool, typically. Xcluder Rodent and Pest Defense makes "cut-and-stuff" stainless steel fiber products. In a seven-day NWRC test, none of 10 Xcluder-blocked gaps were breached by rats and mice attempting to reach their preferred "lure foods" (peanut butter oatmeal balls and, for the house mice, hot dogs and cheese).

BIBLIOGRAPHY

A Quick Word of Introduction

Evans, E. P. *The Criminal Prosecution and Capital Punishment of Animals.* New York: E. P. Dutton and Company, 1906.

1. Maul Cops

Conover, Michael R. *Resolving Human Wildlife Conflicts: The Science of Wildlife Damage Management.* Boca Raton: Lewis Publishers, 2002. Table 3.1: Studies of Nonfatal and Fatal Injuries to Humans by Wildlife in Different Parts of the U.S. and Canada.

Floyd, Timothy. "Bear-Inflicted Human Injury and Fatality." *Wilderness and Environmental Medicine* 10 (1999): 75–87.

U.S. Consumer Product Safety Commission. "Product Instability or Tip-Over Injuries and Fatalities Associated with

Televisions, Furniture, and Appliances: 2012 Report." Graph, p. 17.

Young, Stanley Paul, and Edward Alphonso Goldman. *The Puma.* Washington, DC: American Wildlife Institute, 1946.

2. Breaking and Entering and Eating

Alldredge, Mat W., et al. "Evaluation of Translocation of Black Bears Involved in Human-Bear Conflicts in South-Central Colorado." *Wildlife Society Bulletin* 39, no. 2 (June 2015): 334–40.

Beckmann, Jon P., Carl W. Lackey, and Joel Berger. "Evaluation of Deterrent Techniques and Dogs to Alter Behavior of 'Nuisance' Black Bears." *Wildlife Society Bulletin* 32, no. 4 (2004): 1141–46.

Breck, Stewart W. "Selective Foraging for Anthropogenic Resources by Black Bears: Minivans in Yosemite National Park." *Journal of Mammalogy* 90, no. 5 October 2009): 1041–44.

George, Kelly A., et al. "Changes in Attitude Toward Animals in the United States from 1978 to 2014." *Biological Conservation* 201 (2016): 237–42.

Johnson, Heather E., et al. "Human Development and Climate Affect Hibernation in a Large Carnivore with Implications for

Human-Carnivore Conflicts." *Journal of Applied Ecology* 55, no. 2 (March 2018): 663–72.

Johnson, Heather E., et al. "Assessing Ecological and Social Outcomes of a Bear-Proofing Experiment." *Journal of Wildlife Management* 82, no. 6 (2018): 1102–14.

Linnell, John D. C., et al. "Translocation of Carnivores as a Method for Managing Problem Animals: A Review." *Biodiversity and Conservation* 6, no. 9 (September 1997): 1245–57.

Manning, Elizabeth. "Tasers for Moose and Bears: Alaska Explores Law Enforcement Tool for Wildlife." *Alaska Fish & Wildlife News,* March 2010.

Nelson, Ralph A., et al. "Behavior, Biochemistry, and Hibernation in Black, Grizzly, and Polar Bears." *Proceedings of the International Conference on Bear Research and Management* 5 (1983): 284–90.

Roenigk, Adolph. *Pioneer History of Kansas.* Transcribed by his great-grandniece L. Ann Bowler. Denver, CO, 1933. https://www.kancoll.org/books/roenigk/index.html.

Rogers, Lynn L. "Homing by Radio-Collared Black Bears, *Ursus americanus,* in Minnesota." *Canadian Field Naturalist* 100, January 1986.

Spencer, Rocky D., Richard A. Beausoleil, and Donald A. Martorello. "How Agencies Respond to Human–Black Bear Conflicts: A Survey of Wildlife Agencies in North America." *Ursus* 18, no. 2 (2007): 217–29.

3. The Elephant in the Room

The Asian Elephant (Elephas maximus) of Nagaland: Landscape & Human-Elephant Conflict Management. Dimapur, Nagaland: Government of Nagaland, Wildlife Wing, Department of Forests, Environment and Wildlife.

Gopalakrishnan, Shankar, Terpan Singh Chauhan, and M. S. Selvaraj. "It Is Not Just About Fences: Dynamics of Human-Wildlife Conflict in Tamil Nadu and Uttarakhand." *Economic & Political Weekly* 52: 97–104.

Hindustan Times. "Appetite for Money: Elephants Who Entered a Shop Gorge on Rs 2,000, 500 Notes." April 25, 2017.

———. "Drunken Man Challenges Elephants' Herd, Trampled to Death in Jharkhand." December 19, 2018.

Jayewardene, Jayantha. *The Elephant in Sri*

Lanka. Colombo, Sri Lanka: Wildlife Heritage Trust of Sri Lanka, 1994.

Lahiri-Choudhury, Dhriti K. "History of Elephants in Captivity in India and Their Use: An Overview." *Gajah* 14 (June 1995): 28–31.

McKay, George M. *Behavior and Ecology of the Asiatic Elephant* (Smithsonian Contributions to Zoology, Number 125). Washington, DC: Smithsonian Institution Press, 1973.

Naha, Dipanjan, et al. "Assessment and Prediction of Spatial Patterns of Human-Elephant Conflicts in Changing Land Cover Scenarios of a Human-Dominated Landscape in North Bengal." *PLOS ONE,* February 1, 2019.

Outlook India. "928 Elephants Died Unnaturally Since 2009 Including 565 Due to Electrocution Alone." March 25, 2019.

———. "Delhi Planning to Club Old Age Home with Cow Shelter." January 9, 2019.

Siegel, Ronald K., and Mark Brodie. "Alcohol Self-Administration by Elephants." *Bulletin of the Psychonomic Society* 22, no. 1 (1984): 49–52.

U.S. House of Representatives, Committee on the Judiciary, Hearing before the Subcommittee on Crime. *Captive Elephant*

Accident Prevention Act of 1999. 106th Cong., 2d sess., 2000. H.R. 2273.

4. A Spot of Trouble

Athreya, Vidya. "Is Relocation a Viable Option for Unwanted Animals? The Case of the Leopard in India." *Conservation and Society* 4, no. 3 (2006): 419–23.

Athreya, Vidya, et al. "Translocation as a Tool for Mitigating Conflict with Leopards in Human-Dominated Landscapes of India." *Conservation Biology* 25, no. 1 (November 2010): 133–41.

Corbett, Jim. *The Man-Eating Leopard of Rudraprayag.* New Delhi: Rupa, 2016.

Naha, Dipanjan, S. Sathyakumar, and G. S. Rawat. "Understanding Drivers of Human-Leopard Conflicts in the Indian Himalayan Region: Spatio-Temporal Patterns of Conflicts and Perception of Local Communities Towards Conserving Large Carnivores." *PLOS ONE,* October 2018.

Singh, H. S. *Leopards in the Changing Landscapes.* Dehra Dun: Bishen Singh Mahendra Pal Singh, 2014.

Times of India. "Leopard Enters Hema Malini's House." May 28, 2011.

5. The Monkey Fix

Chauhan, Arvind. "Monkey Snatches Baby from Mom, Kills It." *Times of India,* November 14, 2018.

———. "UP: After Infant's Death, 2 More Toddlers Attacked by Monkeys." *Times of India,* November 17, 2018.

Colagross-Schouten, A., et al. "The Contraceptive Efficacy of Intravas Injection of Vasalgel™ for Adult Male Rhesus Monkeys." *Basic Clinical Andrology* 27, no. 1 (2017), article no. 4.

Gandhiok, Jasjeev, and Paras Singh. "Delhi: Simians Wreak Havoc; Forest Department, Corporations Pass Buck." *Times of India,* January 19, 2019.

Harris, Gardiner. "Indians Feed the Monkeys, Which Bite the Hand." *New York Times,* May 22, 2012.

Killian, G., D. Wagner, and L. Miller. "Observations on the Use of the GnRH Vaccine Gonacon™ in Male White-Tailed Deer (*Odocoileus virginianus*)." *Proceedings of the 11th Wildlife Damage Management Conference,* 2005.

Miller, Lowell A., Kathleen A. Fagerstone, and Douglas C. Eckery. "Twenty Years of Immunocontraceptive Research: Lessons

Learned." *Journal of Zoo and Wildlife Medicine* 44, Supplement 4 (December 2013): S84–S96.

Mohan, Vishwa. "Order to Cull HP's 'Vermin' Monkeys Draws Activists' Ire." *Times of India,* July 19, 2019.

Mohapatra, Bijayeeni, et al. "Snakebite Mortality in India: A Nationally Representative Mortality Survey." *PLOS Neglected Tropical Diseases* 5, no 4 (April 2011): e1018.

Singh, Paras. "Delhi: South Corporation Finally Nets Eight Monkey Catchers." *Times of India,* October 8, 2018.

Times of India. "Teen Killed in Monkey Attack in Kasganj; 5th Death in a Month." December 3, 2018.

———. "Simians Lay Siege to Agra." November 16, 2018.

———. "70-Year-Old Allegedly Stoned to Death by Monkeys; Kin Demands FIR." October 20, 2018.

6. Mercurial Cougars

Beier, Paul, Seth P. D. Riley, and Raymond M. Sauvajot. "Mountain Lions." In *Urban Carnivores: Ecology, Conflict, and Conservation,* edited by Stanley D. Gehrt, Seth P. D. Riley, and Brian L Cypher.

Baltimore: Johns Hopkins University Press, 2010.

Brewster, R. Kyle, et al. "Do You Hear What I Hear? Human Perception of Coyote Group Size." *Human–Wildlife Interaction* 11, no. 2 (Fall 2017): 167–74.

Clemente, Jonathan D. "CIA's Medical and Psychological Analysis Center (MPAC) and the Health of Foreign Leaders." *International Journal of Intelligence and Counterintelligence* 19, no. 3 (2006): 385–423.

Fisher, A. K. "The Hawks and Owls of the United States in Their Relation to Agriculture." Washington, DC: U.S. Department of Agriculture, Division of Ornithology and Mammalogy, Bulletin No. 3, 1893.

Hunter, J. S. "The Mountain Lion." Article manuscript, undated. Joseph S. Hunter Papers, F3735:618. California State Archives: Records of the Division of Fish and Game. [Jay Bruce statement]

"Mountain Lion." Letter from Jay C. Bruce to J. S. Hunter, March 23, 1941. Joseph S. Hunter Papers, F3735:618. California State Archives: Records of the Division of Fish and Game.

Peirce, E. R. "A Method of Determining the Prevalence of Rats in Ships." *The Medical Officer* 43 (1930): 222–24.

Todd, Kim. "Coyote Tracker." *Bay Nature,*

January–March 2018.

Welch, David. "Dung Properties and Defecation Characteristics in Some Scottish Herbivores, with an Evaluation of the Dung-Volume Method of Assessing Occupance." *Acta Theriologica* 27, no. 15 (October 1982): 191–212.

Yiakoulaki, M. D., and A. S. Nastis. "A Modified Faecal Harness for Grazing Goats on Mediterranean Shrublands." *Journal of Range Management* 51, no. 5 (September 1998): 545–46.

Young, Stanley Paul, and Edward Alphonso Goldman. *The Puma.* Washington, DC: The American Wildlife Institute, 1946.

7. When the Wood Comes Down

BC Parks. Wildlife/Danger Tree Assessor's Course Workbook. Revised edition, March 2012. https://www2.gov.bc.ca/assets/gov/environment/plants-animals-and-ecosystems/conservation-habitat-management/wildlife-conservation/wildlife-tree-committee/parks-handbook.pdf.

Brookes, Andrew. "Preventing Death and Serious Injury from Falling Trees and Branches." *Australian Journal of Outdoor Education* 11, no. 2 (2007): 50–59.

Mulford, J. S., H. Oberli, and S. Tovosia. "Coconut Palm-Related Injuries in the Pacific Islands." *ANZ Journal of Surgery* 71, no. 1 (2001): 32–34.

Oregon Fatality Assessment and Control Evaluation. *Fallers Logging Safety* (manual). 2007. https://www.ohsu.edu/sites/default/files/2019-02/ORFACE-Safety Booklet-FallersLoggingSafety-Eng.pdf.

Schmidlin, Thomas. "Human Fatalities from Wind-Related Tree Failures in the United States, 1995–2007." *Natural Hazards* 50, no. 1 (2009): 13–25.

Tribun-Bali. "Falling Durian Possibly Killed Man in West Bali" (via Google Translate). January 28, 2015.

———. "To the Durian Garden Without Head Shield, Kusman Found Dead" (via Google Translate). March 26, 2015.

Walsh, Raoul A., and Lara Ryan. "Hospital Admissions in the Hunter Region from Trees and Other Falling Objects, 2008–2012." *Australian and New Zealand Journal of Public Health* 41, no. 2 (2017): 121–24.

8. The Terror Beans

Arianti, V. "Biological Terrorism in Indonesia." *The Diplomat,* November 20, 2019.

https://thediplomat.com/2019/11/biologi cal-terrorism-in-indonesia/.

"Compound 1080 — Powerful New Rat Killer." Press release, n.d. Fort Collins, CO: National Wildlife Research Center Archive.

Dymock, William, C. J. H. Warden, and David Hooper. *Pharmacographia Indica: A History of the Principal Drugs of Vegetable Origin, Met with in British India.* London: Kegan Paul, Trench, Trübner & Co., 1891.

Eisemann, John D., Patricia A. Pipas, and John L. Cummings. "Acute and Chronic Toxicity of Compound DRC-1339 (3-Chloro-4-Methylaniline Hydrochloride) to Birds." *USDA National Wildlife Research Center Staff Publications* 211, November 2003.

Filmer, Ann. "Safe and Poisonous Garden Plants." University of California, Davis, October 2012. https://ucanr.edu/sites/ poisonous_safe_plants/files/154528.pdf.

Jacobsen, W. C., and S. V. Christierson, eds., Rodent Control Division. "California Ground Squirrels: A Bulletin Dealing with Life Histories, Habits and Control of the Ground Squirrels in California." *Monthly Bulletin of the California State Commission of Horticulture* VII, nos. 11 and 12, November–December 1918.

Jain, Ankita, et al. "Foreign Body (Kidney Beans) in Urinary Bladder: An Unusual Case Report." *Annals of Medicine and Surgery* 32 (August 2018): 22–25.

Karthikeyan, Aishwarya, and S. Deepak Amalnath. "*Abrus precatorius* Poisoning: A Retrospective Study of 112 Patients." *Indian Journal of Critical Care* 21, no. 4 (April 2017): 224–25.

Linz, George M., and H. Jeffrey Honan. "Tracing the History of Blackbird Research Through an Industry's Looking Glass: *The Sunflower Magazine.*" *Proceedings of the 18th Vertebrate Pest Conference,* 1998.

Malik, Balwant Singh. "Punishment of Transportation for Life." *Journal of the Indian Law Institute* 36, no. 1 (1994): 111–20.

Nicholson, Blake. "Debate Rises over Blackbirds." *Bismarck Tribune,* March 18, 2007.

Ogawa, Haruko, and Kimie Date. "The 'White Kidney Bean Incident' in Japan." In *Lectins: Methods and Protocols.* Part of Methods in Molecular Biology book series, volume 1200. New York: Humana Press, 2014.

Ormsbee, R. A. "A Summary of Field Reports on 1080 (Sodium Fluoroacetate)." National Research Council, Insect

Control Committee, Technical Report No. 163. December 17, 1945.

Pincus, Seth H., et al. "Passive and Active Vaccination Strategies to Prevent Ricin Poisoning." *Toxins* 3, no. 9 (September 2011): 1163–84.

Pitschmann, Vladimír, and Zdeněk Hon. "Military Importance of Natural Toxins and Their Analogs." *Molecules* 21, no. 5 (April 2016): 556–78.

Renshaw, Birdsey, to Dr. W. R. Kirner. Memorandum regarding "Animal Poisons" sent from the Office for Emergency Management National Defense Research Committee of the Office of Scientific Research and Development. Washington, DC (1530 P Street, NW), December 30, 1943. Fort Collins, CO: National Wildlife Research Center Archive.

Roxas-Duncan, Virginia I., and Leonard A. Smith. "Of Beans and Beads: Ricin and Abrin in Bioterrorism and Biocrime." *Journal of Bioterrorism & Biodefense* S2:002, January 2012.

Smith, George, and Dick Destiny. "Great WMD Failures: Casey the Castor Oil Killer." *The Register,* October 18, 2006. http://www.theregister.com/2006/10/18/dd_castor_oil_wmd/.

The Sunflower. "Blackbird Project Focuses

on Population Reduction." December 1, 1996.

Thornton, S. L., et al. "Castor Bean Seed Ingestions: A State-Wide Poison Control System's Experience." *Clinical Toxicology* 52, no. 4 (March 2014): 265–68.

Ward, Justus C. "Rodent Control with 1080, ANTU, and Other War-Developed Toxic Agents." *American Journal of Public Health* 36, no. 12 (December 1946): 1427–31.

Wildlife Research Laboratory, Division of Wildlife Research, U.S. Fish & Wildlife Services. "Compound 1080 — A New Agent for the Control of Noxious Mammals." Denver, CO, n.d. Fort Collins, CO: National Wildlife Research Center Archive.

9. Okay, Boomer

Blackwell, Bradley F., Eric Huszar, George M. Linz, and Richard A. Dolbeer. "Lethal Control of Red-Winged Blackbirds to Manage Damage to Sunflower: An Economic Evaluation." *Journal of Wildlife Management* 67, no. 4 (October 2003): 818–28.

Daily News (Perth). "Emus Outwit Gunners." November 4, 1932, p. 1.

————. "Campion Evacuated: Emus Flourish Unharried." November 10, 1932, p. 5.

Daily Telegraph (Sydney). "Not Easy to Kill Emus: A Thousand Rounds Fired, 12 Dead." November 5, 1932, p. 3.

Fisher, Harvey I. "Airplane-Albatross Collisions on Midway Atoll." *The Condor* 68 (May 1966): 229–42.

Flying Magazine. "MATS Versus the Gooney Bird." September 1958.

Frings, Hubert. *The Scientific Scobberlotching of Hubert and Mable Frings.* BTcurlew Press, 2015.

Frings, Hubert, and Mable Frings. "Problems of Albatrosses and Men on Midway Islands." *The Elepaio: Journal of the Hawaii Audubon Society* 20, no. 5 (1959).

Kenyon, Karl W., Dale W. Rice, Chandler S. Robbins, and John W. Aldrich. "Birds and Aircraft on Midway Islands: 1956–57 Investigations." *Special Scientific Report — Wildlife* No. 38. Washington, DC: United States Department of the Interior, Fish and Wildlife Service, January 1958.

Mail (Adelaide). "Request to Use Bombs to Kill Emus." July 3, 1943, p. 12.

Rice, Dale W. "Birds and Aircraft on Midway Islands: 1957–58 Investigations." *Special Scientific Report — Wildlife* No. 44.

Washington, DC: United States Department of the Interior, Fish and Wildlife Service, 1959.

Robbins, Chandler S. "Birds and Aircraft on Midway Islands: 1959–63 Investigations. *Special Scientific Report — Wildlife* No. 85. Washington, DC: United States Department of the Interior, Fish and Wildlife Service, 1966.

Sydney Morning Herald. "War on Emus." October 12, 1932, p. 11.

USFWS — Pacific Region. "Night Vision Trail Cameras Capture Mouse Attacks on Albatross." Video, October 31, 2017.

West Australian (Perth). "War on Emus: Ambush at a Dam." November 8, 1932, p. 8.

Western Mail (Perth). "A Thousand Birds in Luck: Machine Guns Jam." November 10, 1932, p. 28.

10. On the Road Again

Ansari, S. A., et al. "Dorsal Spine Injuries in Saudi Arabia — An Unusual Cause." *Surgical Neurology* 56, no. 3 (2001): 181–84.

Biondi, Kristin M. "White-Tailed Deer Incidents with U.S. Civil Aircraft." *Wildlife Society Bulletin* 35, no. 3 (September

2011): 303–9.

Blackwell, Bradley F., and Thomas W. Seamans. "Enhancing the Perceived Threat of Vehicle Approach to Deer." *Journal of Wildlife Management* 73, no. 1 (2009): 128–35.

Cohen, Bradley S., et al. "Behavioral Measure of the Light-Adapted Visual Sensitivity of the White-Tailed Deer." *Wildlife Society Bulletin* 38, no. 3 (September 2014): 480–85.

D'Angelo, Gino, et al. "Development and Evaluation of Devices Designed to Minimize Deer-Vehicle Collisions." Final Project Report, Daniel B. Warnell School of Forestry and Natural Resources. July 2, 2007.

DeVault, Travis, et al. "Effects of Vehicle Speed on Flight Initiation by Turkey Vultures: Implications for Bird-Vehicle Collisions." *PLOS ONE,* February 4, 2014.

———. "Speed Kills: Ineffective Avian Escape Responses to Oncoming Vehicles." *Proceedings of the Royal Society B: Biological Sciences,* February 22, 2015.

DeVault, Travis, Bradley F. Blackwell, and Jerrold L. Belant, eds. *Wildlife in Airport Environments: Preventing Animal-Aircraft Collisions through Science-Based Management.* Baltimore: Johns Hopkins University Press, 2013.

DeVault, Travis, Thomas W. Seamans, and Brad Blackwell. "Frontal Vehicle Illumination via Rear-Facing Lighting Reduces Potential for Collisions with White-Tailed Deer." *Ecosphere* (manuscript accepted).

Dolbeer, Richard A., et al. "Wildlife Strikes to Civil Aircraft in the United States, 1990–2015." *National Wildlife Strike Database Serial Report Number 21,* July 2015.

Gens, Magnus. "Moose Crash Test Dummy." Master's thesis, Royal Institute of Technology, Stockholm, Sweden. *VTI särtryck* 342, 2001.

Hughson, Debra L., and Neal Darby. "Desert Tortoise Road Mortality in Mojave National Preserve, California." *California Fish and Game* 99, no. 4 (September 2013): 222–32.

Kennedy Space Center Status Report. "Roadkill Roundup." April 26, 2006.

Kim, Sharon, and A. Robertson Harrop. "Maxillofacial Injuries in Moose–Motor Vehicle Collisions Versus Other High-Speed Motor Vehicle Collisions." *Canadian Journal of Plastic Surgery* 13, no. 4 (December 2005): 191–94.

Knapp, Keith K. "Deer-Vehicle Crash Countermeasure Toolbox." Iowa State

University Institute for Transportation, Deer-Vehicle Crash Information Clearinghouse, 2001.

Pynn, Tania P., and Bruce R. Pynn. "Moose and Other Large Animal Wildlife Collisions: Implications for Prevention and Emergency Care." *Journal of Emergency Nursing* 30, no. 6 (2004): 542–47.

Riginos, Corinna, et al. "Wildlife Warning Reflectors and White Canvas Reduce Deer-Vehicle Collisions and Risky Road-Crossing Behavior." *Wildlife Society Bulletin* 42, no. 4 (March 2018): 1–11.

Al-Sebai, M. W., and S. Al-Zahrani. "Cervical Spinal Injuries Caused by Collision of Cars with Camels." *Injury* 28, no. 3 (April 1997): 191–94.

Simmons, James Raymond. *Feathers and Fur on the Turnpike.* Boston: The Christopher Publishing House, 1938.

Williams, Allan F., and Joann K. Wells. "Characteristics of Vehicle-Animal Crashes in Which Vehicle Occupants Are Killed." *Traffic Injury Prevention* 6, no. 1 (2005): 56–59.

11. To Scare a Thief

Avery, Michael L., et al. "Dispersing Vulture Roosts on Communication Towers." *Jour-*

nal of Raptor Research 36, no. 1 (February 2002): 45–50.

Bildstein, Keith. *Raptors: The Curious Nature of Diurnal Birds of Prey.* Ithaca, NY: Cornell University Press, 2017.

Blackwell, Bradley, Thomas W. Seamans, Morgan B. Pfeiffer, and Bruce N. Buckingham. "European Starling (*Sturnus vulgaris*) Reproduction Undeterred by Predator Scent Inside Nest Boxes." *Canadian Journal of Zoology* 96, no. 9 (2018): 980–86.

Chipman, Richard B., et al. "Emergency Wildlife Management Response to Protect Evidence Associated with the Terrorist Attack on the World Trade Center, New York City." *Proceedings of the 21st Vertebrate Pest Conference,* 2004.

Mauldin, Richard E., et al. "Development of a Synthetic Materials Mimic for Vulture Olfaction Research." *Proceedings of the 10th Damage Management Conference,* 2003.

Seamans, Thomas W. "Response of Roosting Turkey Vultures to a Vulture Effigy." *Ohio Journal of Science* 104, no. 5 (December 2004): 136–38.

Tillman, Eric A., John S. Humphrey, and Michael L. Avery. "Use of Vulture Carcasses and Effigies to Reduce Vulture Damage to Property and Agriculture." *Proceedings of the 20th Vertebrate Pest Conference,* 2002, pp. 123–28.

12. The Gulls of St. Peter's

Glahn, James F., Greg Ellis, Paul Fioranelli, and Brian Dorr. "Evaluation of Moderate and Low-Powered Lasers for Dispersing Double-Crested Cormorants from Their Night Roosts." *Proceedings of the Ninth Wildlife Damage Management Conference,* January 2001.

Glatz, Carol. "Feathery Fiascos: The Unfortunate Prey for Peace." Catholic News Service Blog, January 27, 2014.

Graham, Frank, Jr. *Gulls: An Ecological History.* New York: Van Nostrand Reinhold, 1975.

Linton, E., et al. "Retinal Burns from Laser Pointers: A Risk in Children with Behavioral Problems." *Eye* 33, no. 3 (March 2019): 492–504.

Markham, Gervase. *Markham's Farewell to Husbandry.* London: Nicholas Oakes for John Harrison, 1631.

Parsons, Jasper. "Cannibalism in Herring Gulls." British Birds (newsletter), December 1, 1971.

Vickery, Juliet A., and Ronald W. Summers. "Cost-Effectiveness of Scaring Brent Geese (*Branta b. bernicla*) from Fields of Arable Crops by a Human Bird Scarer." *Crop Protection* 11, no. 5 (October 1992): 480–84.

13. The Jesuit and the Rat

Francis (pope). " 'Laudato Si": On Care for Our Common Home." Encyclical of the Holy Father on Climate Change and Inequality. http://w2.vatican.va/content/francesco/en/encyclicals/documents/papa-francesco_20150524_enciclica-laudato-si.html.

Philippi, Dieter. "Campagi — The Footwear of the Pope and the Clergy." http://www.dieter-philippi.de/en/ecclesiastical-fineries/campagi-the-footgear-of-the-pope-and-the-clergy.

14. Killing with Kindness

Adams, Lowell W., J. Hadidian, and V. Flyger. "Movement and Mortality of Translocated Urban-Suburban Grey Squirrels." *Animal Welfare* 13, no. 1 (February 2004): 45–50.

American Veterinary Medical Association.

AVMA Guidelines for the Euthanasia of Animals. American Veterinary Medical Association: 2013 edition.

————. "AVMA May Change Guidance for CO2 Euthanasia in Rodents." *JAVMA News,* January 1, 2019.

Egerton, Rachael. " 'Unconquerable Enemy or Bountiful Resource?' A New Perspective on the Rabbit in Central Otago." Bachelor's thesis, University of Otago, Dunedin, New Zealand, March 18, 2014. Australian & New Zealand Environmental History Network, https://www.environmentalhistory-au-nz.org/publications/.

King, Carolyn M. "Liberation and Spread of Stoats (*Mustela erminea*) and Weasels (*M. nivalis*) in New Zealand, 1883–1920." *New Zealand Journal of Ecology* 41, no. 2 (2017): 163–76.

Littin, Kate E., et al. "Behavior and Time to Unconsciousness of Brushtail Possums (*Trichosurus vulpecula*) After a Lethal or Sublethal Dose of 1080." *Wildlife Research* 36, no. 8 (2009): 709–20.

Mason, G., and K. E. Littin. "The Humaneness of Rodent Pest Control." *Animal Welfare* 12, no. 1 (February 2003): 1–37.

Morriss, Grant A., Graham Nugent, and Jackie Whitford. "Dead Birds Found After

Aerial Poisoning Operations Targeting Small Mammal Pests in New Zealand 2003–14." *New Zealand Journal of Ecology* 40, no. 3 (January 2016): 361–70.

Robinson, Weldon B. "The 'Humane Coyote-Getter' vs. the Steel Trap in Control of Predatory Animals." *Journal of Wildlife Management* 7, no. 2 (April 1943): 179–89.

Stats NZ. "Conservation Status of Indigenous Land Species." April 17, 2019. https://www.stats.govt.nz/indicators/conservation-status-of-indigenous-land-species.

Warburton, Bruce, Nick Poutu, and Ian Domigan. "Effectiveness of the Victor Snapback Trap for Killing Stoats." *DOC Science Internal Series* 83. Wellington: New Zealand Department of Conservation. October 2002.

Warburton, Bruce, Neville G. Gregory, and Grant Morriss. "Effect of Jaw Shape in Kill-Traps on Time to Loss of Palpebral Reflexes in Brushtail Possums." *Journal of Wildlife Diseases* 36, no. 1 (2000): 92–96.

15. The Disappearing Mouse

Kimball, Bruce, et al. "Development of Artificial Bait for Brown Treesnake Suppression." *Biological Invasions* 18 (2016): 359–69.

Pitt, William C., et al. "Physical and Behavioral Abilities of Commensal Rodents Related to the Design of Selective Rodenticide Bait Stations." *International Journal of Pest Management* 57, no. 3 (July–September 2011): 189–93.

ABOUT THE AUTHOR

Mary Roach is the author of five best-selling works of nonfiction, including *Grunt, Stiff,* and, most recently, *Fuzz.* Her writing has appeared in *National Geographic* and the *New York Times Magazine,* among other publications. She lives in Oakland, California.